Bruno de Finetti (Ed.)

Economia Matematica

Lectures given at a Summer School of the
Centro Internazionale Matematico Estivo (C.I.M.E.),
held in Frascati (Romo), Italy,
August 22-30, 1966

FONDAZIONE
CIME
ROBERTO CONTI

 Springer

C.I.M.E. Foundation
c/o Dipartimento di Matematica "U. Dini"
Viale Morgagni n. 67/a
50134 Firenze
Italy
cime@math.unifi.it

ISBN 978-3-642-11044-3   e-ISBN: 978-3-642-11045-0
DOI:10.1007/978-3-642-11045-0
Springer Heidelberg Dordrecht London New York

Printed on acid-free paper

Springer.com

CENTRO INTERNAZIONALE MATEMATICO ESTIVO

(C. I. M. E.)

4° Ciclo - Villa Falconieri (Frascati) 22-30 agosto 1966

"ECONOMIA MATEMATICA"

Ciclo diretto dal Prof. B. de FINETTI

finanziato dall'Ente per gli studi monetari,

bancari e finanziari "L. EINAUDI" -

CENTRO INTERNATIONALE MATEMATICO ESTIVO

(C. I. M. E.)

S. N. AFRIAT

" ECONOMIC TRANSFORMATION"

Corso tenuto a Villa Falconieri (Frascati) dal 22 al 30 agosto 1966

Economic Transformation

by

S. N. Afriat

Purdue University

1: Transformation Possibility

Economics is especially concerned with possessions; and the institution which is of greatest importance to economics is the claim, or enforcible right to possession. Theories of economics deal with economic agents, their possible states, and their possible actions affecting states. Economic state is described by possessions; economic action alters state, and can be described as a transformation of possessions; and these are the two essential aspects of economic agents. An action, when it is not just a constraint, is a choice. To give account of a choice, there has to be shown the variety of possibilities, and then the motive for decision. Here to be considered is the structure of the variety of possibilities before an economic agent.

Possessions are described as compositions of goods of various kinds and amounts, in other words as stocks of goods. Possible possessions, or stocks, are thus represented by the vectors in $\Omega = \{x : x = (x_1, \ldots, x_n) \geq 0\}$. A transformation of a given stock $x \epsilon \Omega$ possessed by an agent results in the attainment of possession of some other stock $y \epsilon \Omega$. Thus the possible transformations of the agent are described by a relation $T \subseteq \Omega \times \Omega$, between all possible stocks,

where xTy denotes $(x,y) \in T$ and asserts the possibility of the transformation of x into y; that is, were the agent in the economic state
defined by possession of x, it would be possible, by available means,
to attain the state y. Those means might be the exchanges which take
place between agents, or through markets, or in the input-output of
industrial processes permitted by technology. Whatever the sources
of possibility, they are limited, or the economic meaning of goods would
vanish.

The question now is the structure which is to be assumed for
<u>transformation-possibility relation</u> T. One special structure will
arise from the Koopmans static model of production activity[1]. The
same formal structure will here be established on the basis of six
independent axioms. Five of these virtually are inseparable from the
concept of T, and are rendered true by proper interpretation. The
remaining one (Axiom 3) more has the character of a special
assumption. The essential distinction between Koopman's static model
and von Neumann's dynamic model[2] of production appears especially in
one of the axioms (Ax.1).[3]

2: Koopman's Transformation Model

Koopmans assumes m basic activities A_i, involving n basic goods G_j. The outcome of activity A_i is that a quantity a_{ij} of good G_j is produced, or equivalently $-a_{ij}$ is consumed. Hence the vector a_i with elements a_{ij} represents the outcome of activity A_i. The activities A_i can be combined with any intensities $w_i \geq o$ to form an activity, symbolically denoted $A = \Sigma w_i A_i$, whose outcome is represented by the vector $a = \Sigma w_i a_i$. Accordingly, all the activities, thus generated by the basic activities, form a system $\mathcal{A}$, whose outcomes are represented by the vectors in the convex cone V generated by the vectors a_i.

g.1 ⟩

In order to be able to perform an activity A, it is necessary to possess in sufficient quantities the goods it consumes. That is, considering an agent whose possession of goods is given by x∈Ω, it is necessary that x+a∈Ω, where a∈V is the outcome of the activity A, otherwise some goods would be overexhausted by the activity. Performed in conjunction with possession x, the activity then appears as affecting the transformation of x∈Ω into another possible possession x+a∈Ω. Hence in regard to any two possible possessions x, y∈Ω of an agent who has command of the activities $\mathcal{A}$, a necessary and sufficient condition that there exists an activity A ∈ $\mathcal{A}$ which will effect the transformation of possession from x into y is that y-x∈V. With the system of activities $\mathcal{A}$ as the sole source of means for effecting transformations of possessions, there is established a transformation possibility relation T such that

$$xTy \quad \Leftrightarrow \quad y\text{-}x \in V.$$

Fig. 2 ⟩

That is, if

$$E_V = \{(x,y) : y-x \epsilon V; x,y \epsilon \Omega\},$$

then $T = E_V$.

Thus Koopman's production activity model leads to consideration of a transformation-possibility relation of the form $T = E_V$ where V is a closed convex cone. A finitely generated, and therefore closed cone was obtained from the model; and now a general closed convex cone V can be assumed in the construction of the relation E_V, which can be said to have V as its <u>transformation displacement cone</u>.

Any ordered couple (x,y) of n-vectors x, y can be identified with the 2n-vector $z = (x, y)$ which it represents in partitioned form. Accordingly, any $T \subseteq \Omega \times \Omega$, as a set of ordered couples of n-vectors, is identified with a set in the space of 2n-vectors, and, as such, there is meaning to the assertion that T be a convex 2n-cone. With this understanding, there is the following proposition:

<u>If V is a convex n-cone, then E_V is a convex 2n-cone.</u>

Thus, assume V is a convex cone, that is

$$a_o, a_1 \epsilon V \quad \text{and} \quad \lambda \geq 0 \Rightarrow a_o \lambda, \ a_o + a_1 \epsilon V.$$

It will be shown that E_V is a convex cone, that is

$$(x_o, y_o), \ (x_1, y_1) \ \epsilon \ E_V \ \& \ \lambda \geq 0 \ \Rightarrow \ (x_o, y_o)\lambda, \ (x_o, y_o) + (x_1, y_1) \epsilon \ E_V$$

The hypothesis here is

$$y_o - x_o = a_o, \ y_1 - x_1 = a_1 \ \text{where} \ a_o, \ a_1 \epsilon V,$$

and the conclusion is

$$(x_o\lambda, y_o\lambda), \quad (x_o+x_1, y_o+y_1) \in E_V \ ,$$

that is

$$y_o\lambda - x_o\lambda, \quad (y_o+y_1)-(x_o+x_1) \in V \ ,$$

that is

$$(y_o-x_o)\lambda, \quad (y_o-x_o)+(y_1-x_1) \in V,$$

that is

$$a_o\lambda, \quad a_o+a_1 \in V \ .$$

But this follows from the hypothesis, since V is a convex cone.

However, it is not true that if T is a convex 2n-cone, then it is of the form E_V where V is some convex n-cone, as is obvious.

A relation $T \subseteq \Omega\times\Omega$ may be called translatable if

$$xTy \ \& \ z \geqq 0 \ \Rightarrow \ (x+z)T(y+z).$$

<u>Any relation of the form E_V, where V is any set of vectors, is translatable</u>.

For, if xTy and $z \geq 0$, then x, $y \in \Omega$ so that x+z, $y+z \in \Omega$ and $y-x \in V$ so that $(y+z)-(x+z) \in V$, whence $(x+z)T(y+z)$.

Any relation $T \subseteq \Omega\times\Omega$ is to be called uniform if

$$xTy \ \& \ \lambda \geq 0 \ \Rightarrow \ (x\lambda)T(y\lambda);$$

and it is called reflexive if xTx, and transitive if

$$xTy \ \& \ yTz \ \Rightarrow \ xTz.$$

<u>If V if a convex cone then E_V is uniform, reflexive and transitive.</u>

Thus, assume V is a convex cone. Clearly xE_Vx, since $x-x\epsilon V$, that is $o\epsilon V$, whence E_V is reflexive. If xE_Vy, that is $y-x\epsilon V$, then $(y-x)\lambda\epsilon V$ for $\lambda \geq 0$, that is $y\lambda-x\lambda\epsilon V$, that is $(x\lambda)E_V(y\lambda)$, whence E_V is homogeneous. Again, if xE_Vy and yE_Vz, that is $y-x$, $z-y\epsilon V$, then $(y-x)+(z-y)\epsilon V$, that is $z-x\epsilon V$, that is xE_Vz, whence E_V is transitive.

While if T is a convex cone of 2n-vectors[4] it must be homogeneous, it clearly need not be reflexive or transitive, and therefore, by the proposition just proved, need not be of the form E_V where V is some convex cone of n-vectors, as was remarked previously.

To any convex cone V there corresponds a dual convex cone, defined by

$$U = \{u : u'a \leq 0 \text{ for all } a\epsilon V\},$$

provided this set is non-empty. By the duality theorem[4a] for closed convex cones, the dual of the dual of V then exists, and is again V, that is

$$V = \{a : u'a \leq 0 \text{ for all } u\epsilon U\},$$

or equivalently,

$$a\epsilon V \Leftrightarrow u'a \leq 0 \text{ for all } u\epsilon U.$$

Since

$$xE_Vy \Leftrightarrow y-x\epsilon V,$$

it follows that

$$xE_Vy \Leftrightarrow u'x \geq u'y \text{ for all } u\epsilon U.$$

Thus if

$$I_U = \{(x,y) : u'x \geq u'y \text{ for all } u\epsilon U; \ x,y\epsilon\Omega\}$$

defines a relation between the elements of Ω corresponding to a convex cone U, then

$$I_U = E_V$$

where V is the dual of U.

The formulae E_V, I_U give dual, equivalent forms of definition
of a relation $T \subseteq \Omega \times \Omega$ associated with a cone V and its dual U.
They may be distinguished as the extensional and intensional forms of
definition.[5]

If V is finitely generated then, by a familiar theorem,[5a] so is
its dual U. In this case let $u_1, \ldots, u_k$ denote a set of generators
of U, so

$$U = \{\Sigma u_r \lambda_r : \lambda_r \geq 0\}.$$

Then clearly

$$xTy \Leftrightarrow u_r'x \geq u_r'y \quad \text{for } r=1,\ldots,k ,$$

that is, xTy holds on condition that x, y satisfy a system of k homogeneous linear inequalities. For if these are satisfied, then it
follows that

$$(\Sigma u_r \lambda_r)'x \geq (\Sigma u_r \lambda_r)'y \quad \text{for all } \lambda_r \geq 0,$$

that is,

$$u'x \geq u'y \quad \text{for all } u \epsilon U;$$

and conversely. Hence the following proposition:

If $T = I_U$ where U is finitely generated, then there exists a
finite set of vectors $u_1, \ldots, u_k$ such that

$$xTy \Leftrightarrow u_r'x \geq u_r'y \quad \text{for } r=1,\ldots,k.$$

There are two further properties which are generally going to
be considered in regard to a transformation-possibility relation T.
One, to be called the Axiom of Annihilation, and which means that any
possession can be annihilated, that is, transformed into the null
possession o, which has only zero quantities of goods, is stated xTo.
For a relation of the form $T = E_V$, this requires $o-x \epsilon V$ for all $x \epsilon \Omega$,
that is $-\Omega \subseteq V$. The other condition, to be called the Axiom of
Economy, and which means that in all possible transformations there
is no gain without some loss,[6] is stated $x \leq y \Rightarrow x\bar{T}y$, or what is
the same, $x \leq y$ & $xTy \Rightarrow x=y$. Applied to $T = E_V$, and taking $z=y-x$,
this means that if $z \epsilon \Omega$, that is $z \geq o$, and if $z \epsilon V$ then $z=o$; that is
$\Omega \cap V = 0$, where $0 = \{o\}$. Thus the axioms of annihilation and economy
applied to $T = E_V$ are equivalent to the conditions

$$-\Omega \subseteq V, \quad \Omega \cap V = 0.$$

Fig. 3)

It appears from these that $-\Omega$, V are two convex sets whose interiors
are non-empty and disjoint and therefore, by a general theorem on
convex sets,[6a] are separated by some hyperplane through their intersection, that is through the point o. Accordingly, there exists at
least one vector $u \neq o$ such that $z \epsilon V \Rightarrow u'z \leq o$, and $z \epsilon \Omega \Rightarrow u'z \geq o$,
in which case $u > o$, since $z>o \Rightarrow u'z > o$. It follows that the dual
cone U of the cone V exists, and moreover that $U \subseteq \Omega$.[6b] Thus it appears
that if T is a transformation-possibility relation which satisfies
the axioms of annihilation and of economy, and which is of the form
$T = E_V$, where V is a closed convex cone, then the dual cone U of V
must exist, and be non-negative, and give $T = I_U$.

But any cone $U \subseteq \Omega$ has a dual V, moreover such that $-\Omega \subseteq V$,
$\Omega \cap V = 0$, whence the following now appears:

Given any relation $T \subset \Omega \times \Omega$, it is of the form $T = E_V$ where V is
a closed convex cone, and further it satisfies the Axioms of Annihilation and Economy, if and only if it is of the form $T = I_U$ where U
is a closed convex cone, and further $U \subseteq \Omega$, and in this case U, V
are duals.

So far there has been formulation of the concept of a transformation-possibility relation $T \subseteq \Omega \times \Omega$, with indication of the universality
of its scope in analytical economics. Then Koopmans' static production
activity model was shown to lead to the form $T = E_V$ where V is a closed
convex cone. Then the requirement that T should have this form and
satisfy the Axioms of Annihilation and of Economy led to the form $T = I_U$
where U is a non-negative closed convex cone. This last form is the
one which is important in this investigation, and which is going to
be subjected to an axiomatic analysis. But first von Neumann's dynamic production model will be reviewed; and remark will be made on
the similarity and contrast between the models of Koopmans and von
Neumann, and between static and dynamic transformation-possibility
relations in general.

In conclusion it can be noted that, with $u_1, \ldots, u_k$ as generators
of the dual U of the activity cone V, if $F(z)$ is the non-decreasing
homogeneous convex function given by $F(z) = \max\{u_r z : r = 1, \ldots, k\}$,
then $V = \{z : F(z) \leq 0\}$. Efficient activities are characterized by
the condition $F(z) = 0$.

3: von Neumann's Transformation Model

Transformations which take place through a certain span of time, say some N unit periods, give a transformation-possibility relation T_N defined for that time span.[7] While in the concept of a static transformation-possibility relation T, the transitivity condition xTy & yTz $\Rightarrow$ xTz is inseparable, since a passage from x to y, and then from y to z gives a passage from x to z, by the nature of what is meant. the same is not the ease for a dynamic relation such as T_N. Instead,

$$xT_M y \ \& \ yT_N z \ \Rightarrow \ xT_{M+N}z,$$

that is $T_M T_N \subseteq T_{M+N}$,[8] is the natural property. That is, if there is a passage from x to y through M periods, and from y to z through N periods, then there is, at least, a passage from x to z through M+N periods. The transitivity condition is the essential distinction of a relation T which is independent of time from a dynamic relation such as T_N. This is reflected in the forms of the transformation-possibility relations which arise from the production-activity model of Koopmans, and the production-process model of von Neumann, one dealing with activities without explicit reference to time, and the other with processes applying specifically through a unit time-period.

von Neumann[9] assumes m basic processes P_i involving n basic goods G_j, carried out over a time-span of unit duration. The process P_i transforms possession of quantities $a_{ij} \geq o$ of the goods into possession of quantities $b_{ij} \geq o$, from the beginning to the end of the duration. That is, if a_i, b_i are the n-vectors with elements a_{ij}, b_{ij} then P_i will effect, across a unit time-span, the transformation of stock possessed from a_i at the beginning to b_i at the end. The processes P can be combined with any intensities $w_i \geq o$ to form a process, symbolically denoted $P = \Sigma w_i P_i$, which will transform a stock from $x = \Sigma w_i a_i$ at one moment to $y = \Sigma w_i b_i$ a unit of time later. Thus they establish a transformation-possibility relation $T_1 \subseteq \Omega \times \Omega$ for a unit of time such that

$$x T_1 y \quad \Leftrightarrow \quad x = \Sigma w_i a_i, \; y = \Sigma w_i b_i \quad \text{for some } w_i \geq 0.$$

But $x T_1 y$ is $(x,y) \in T_1$, and the definition of T_1 is restated $T_1 = V_1$ where

$$V_1 = \{ \Sigma (a_i, b_i) w_i : w_i \geq 0 \} ,$$

that is, T_1 is the non-negative convex 2n-cone V_1, the generators of which are the 2n-vectors given in partitioned form by (a_i, b_i). The previously considered static relation $T = E_V$, defined with V a convex n-cone, also had the property of being a 2n-cone, as was shown. But it also had, following from its form, the property of transitivity not shared by T_1.

The Axiom of Disposal, for a static or dynamic relation, is

$$x_o T y_o \; \& \; x_1 \geq x_o \; \& \; y_1 \leq y_o \; \Rightarrow \; x_1 T y_1. {}^{10}$$

For a static relation of the form $T = E_V$ for some convex n-cone V, the Axiom of Annihilation $x T o$ is sufficient, besides being obviously necessary. For a dynamic relation in the form of a non-negative convex 2n-cone $T_1 = V_1$, disposal is not automatic in the form, but can be imposed by redefinition of T_1 as T_1^*, where

$$x T_1^* y = (\forall a, b) \; x \geq a \; \& \; y \leq b \; \& \; a T_1 b .$$

This defines the smallest relation T_1^* which contains T_1 and satisfies the axiom of disposal. It is easy to see if T_1 is a non-negative 2n-cone then so also is T_1^*. A disposal process is not among von Neumann's explicit assumptions, though it does enter implicitly. The condition that T_1 satisfy the axiom of disposal can be stated $T_1^* = T_1$. In case $T_1^* \neq T_1$, a way of enlarging T_1 to T_1^*, keeping the explicit form of the von Neumann model, is to incorporate among the generators of V_1 some further generators, corresponding to further processes, which will introduce disposal as resulting from processes of the system; and this is certainly possible since T_1^* is a cone, and contains T_1. But, as can be seen from von Neumann's account, disposal can also be taken care of implicitly without having it as an explicit process of the system as has been the method of some other writers.

It is supposed that all the von Neumann processes P, generated by combination of the basic processes, constitute a system of processes ρ which, in a unit of time, can effect the transformation of a stock $x \epsilon \Omega$ into a stock $y \epsilon \Omega$, provided $x \geq \Sigma w_i a_i$ and $y \leq \Sigma w_i b_i$ for some $w_i \geq o$. Thus, the condition that, with a given stock $x \epsilon \Omega$, some process P of the system will be feasible is that sufficient quantities be held of the goods required in at least one process $P = \Sigma w_i P_i$ of ρ, that is $x \geq \Sigma w_i a_i$ for some $w_i \geq o$, assuming disposal of the residue $x - \Sigma w_i a_i$ not needed for the process. The outcome of the transformation is then the stock $\Sigma w_i b_i$, or, allowing disposal, any $y \epsilon \Omega$ such that $y \leq \Sigma w_i b_i$. For some process $P^* = \Sigma w_i^* P_i$ to have applicability to this outcome through the next unit of time, the same condition must hold for this outcome, that is $\Sigma w_i b_i \geq \Sigma w_i^* a_i$ for some $w_i^* \geq o$. Then

$P^* = \sum_i w_i^* P_i$ is an applicable process, and a feasible successor to the process P. The compounding of the processes in chains, each linked with a feasible successor, gives the ramification of the system ρ through indefinitely extended time-periods. There is the possibility of extinction at any stage where no feasible successor exists. If the processes incorporate the entire means of perpetuating stock, then the extinction at some stage of all possible chains over N periods, starting from some initial stock, means the annihilation of stock by the Nth period. Generally, there will be an N-period transformation-possibility relation $T_N \subseteq \Omega \times \Omega$ established by the system of processes ρ, where $x T_N y$ means there exists a feasible chain of N processes, the first one applicable to x, and the final one resulting in y; and, as remarked, there is the possibility that $x T_N y \Rightarrow y = o$.

The relation $T_N = T_1^N$, where T_1^N denotes the product of N replicas of T_1, has its structure implicit in the structure of T_1, and there are various properties which, when possessed by T_1, must necessarily be shared by T_N. If T_1 is a non-negative convex 2n-cone then so is T_N. If T_1 is transitive, so is T_N, and moreover $T_N \subseteq T_1$; and in case T_1 is also reflexive, so is T_N, and $T_N = T_1$. But transitivity is not a usual property of dynamic relations. However, the relation $T = T_1 \cup T_2 \cup \ldots$ (to ∞) is necessarily transitive, from its form; it gives the transformations which are possible, regardless of time-duration, and some of the same properties are fitting for it as for a static transformation-possibility relation.

von Neumann considered the existence of a process $P = \sum_i w_i P_i$ admitting a growth factor $\alpha > o$, that is such that the process

$P* = \alpha P$ ($=\Sigma \alpha w_i P_i$) is a feasible successor of P, that is

$$\alpha \Sigma_i w_i a_i \leq \Sigma_i w_i b_i \; ;$$

in which case it follows that $P*$ also admits α as a growth factor, and that $P** = \alpha P* = \alpha^2 P$ is a feasible successor of $P*$, and admits α as a growth factor, and so on indefinitely. If $P = \Sigma_i w_i P_i$ is any process admitting a growth factor α and if $a = \Sigma_i w_i a_i$, then

$$x \geq a \ \& \ o \leq y \leq \alpha^N a \ \Rightarrow \ x T_N y.$$

The existence of processes admitting a positive growth factor α is of special interest, since with them, and starting with a suitable stock, the system can guarantee avoiding certain annihilation after a sufficient but finite lapse of time, even if $\alpha < 1$, and the system may run down towards annihilation indefinitely, but without ever actually attaining it.

von Neumann's investigation shows conditions for the existence of processes admitting a growth factor, and also considers a dual problem involving prices and an interest factor, and "equilibrium" conditions, involving the dual pair of problems, which result in growth and interest factors being determined equal, one at a maximum and the other at a minimum.

Some other properties can be considered as bearing on transformations, static or dynamic. The reflexivity condition xTx is important for both static and dynamic relations, but with different significance. As a property it is inseparable from the static case, as is going to be proposed later. For dynamic relations, as the Axiom of Storage, it means the possibility of any stock or part of a stock held being perpetuated unchanged through an indefinite time. In contrast with the static case, it here has the character of a special assumption. For the Axiom of Separation, xTy & $z \geq o \Rightarrow (x+z)T(y+z)$, similar remarks can be made; and it implies reflexivity if it is allowed that oTo, as must be allowed if certain other axioms are assumed, such as, most simply, the Axiom of Annihilation xTo.

4: Axioms of Transformation-Possibility

The following axioms will be considered in regard to a transformation-possibility relation $T \subseteq \Omega \times \Omega$.

1. Axiom of Procession: xTy & yTz ⇒ xTz

That is xTyTz ⇒ xTz, equivalently xTyT...Tz ⇒ xTz, in other words T is a transitive relation. Two transformations, of x to y and y to z, in that order, with the final stock of the first identical with the initial stock of the succeeding one, can be said to be linked, and to have as their result the transformation from x to z, with y as intermediate stage. A chain of transformations x to y and y to ... to z, each linked to its successor, defines a procession of transformations, with the transformation of x to z as its result, with y, ... as intermediate stages. The axiom states that the result of a procession of possible transformations is a possible transformation. This axiom is inseparable from the concept of transformation.

2. Axiom of Separation: xTy & z ≥ o ⇒ (x+z)T(y+z)

The transformation of a stock x+z into a stock y+z is possible, by the separate transformation of the part x of the one into the part y of the other, if this transformation is possible when these parts are entire possessions, and there is identity between the residual parts. Thus, x is part of a stock x+z, with z as residual. With x as an entire possession the transformation of x into y is possible. Then, according to the axiom, by separate transformation of x in the total x+z, so that x can still become y without obstruction from the presence of z, the transformation of x+z into y+z is possible. Imagine

Fig.5〉 a heap of goods, separate it into two parts, modify one part, and put the two heaps together again. The axiom asserts that the possible

modifications of the one part are the same as when it is alone, and
without any possible interference from the presence of the other part.
With certain interpretations this axiom can appear objectionable. It
is possible that the mere presence of z, say in a factory, fills a
capacity, perhaps just space, which is essential to carry out certain
transformations of x which are possible when x is alone. There is an
error in formulation here, in that the capacity is itself a stock of a
good which must be put among the other goods in the heap. Were
stocks formulated as pure claims to stocks, rather than as material
possessions which, in the taking, necessarily fill certain capacities
already possessed, that is, were possessions considered as becoming
actual by the transformation of pure claims, such objections could not
arise. It is illegal to park an automobile on the streets. To have
possession of one reguires a garage. A pure claim to an automobile
and an unoccupied garage has the possibility of being transformed into
the possession of an automobile and an occupied garage. Distinction
between pure claim and material possession gives a framework for in-
terpreting this axiom as having inevitability.

The Axioms of Procession and Separation together have the following consequence:
$$x_0 T y_0 \ \& \ x_1 T y_1 \ \Rightarrow \ (x_0 + x_1) T (y_0 + y_1),$$
which can be called the Axiom of Additivity. Thus, by separation,
$x_0 T y_0$ implies $(x_0 + x_1) T (y_0 + x_1)$ and $x_1 T y_1$ implies $(x_1 + y_0) T (y_1 + y_0)$, and
these conclusions, by procession, together imply $(x_0 + x_1) T (y_0 + y_1)$.
An extended, equivalent form of the additivity axiom is
$$x_1 T y_1 \ \& \ \ldots \ \& \ x_m T y_m \ \Rightarrow \ (x_1 + \ldots + x_m) T (y_1 + \ldots + y_m),$$

as appears by induction on m. A particular consequence is

$$xTy \;\Rightarrow\; (mx)T(my), \; m = 1,2,\ldots,$$

$Fig.6$ which condition can be called the Axiom of Multiplication. With this

let there also be considered an Axiom of Division, defined by

$$xTy \;=\; (\tfrac{1}{m}x)T(\tfrac{1}{m}y), \; m = 1,2,\ldots$$

The conjunction of the Axioms of Multiplication and Division is ob-

viously equivalent to the condition

$$xTy \;\Rightarrow\; (\tfrac{p}{q}x)T(\tfrac{p}{q}y), \; p,q = 1,2,\ldots$$

which can be called the Axiom of Rational Similarity. It is a case of

the Axiom of Similarity which is going to be considered next. But, on

assumption of the Axiom of Continuity, which is going to be considered

last, the Axioms of Similarity and Rational Similarity are equivalent,

as will be shown.

It appears thus that, on condition of the Axioms of Procession,

Separation and Continuity, the Axiom of Division is equivalent to the

Axiom of Similarity about to be considered, even though, without that

condition, it is properly weaker. Thus, given that condition, the

Axiom of Division shows the essential further condition, contained in

the Axiom of Similarity, and which is imposed by its conjunction.

Nevertheless this axiom needs to be considered also in isolation, as

follows.

3. Axiom of Similarity: $xTy \;\&\; \lambda \geq o \;\Rightarrow\; (x\lambda)T(y\lambda)$

Two transformations of the form x to y and $x\lambda$ to $y\lambda$, where

$x,y \epsilon \Omega$ and $\lambda \geq o$, are said to be similar, and to differ just in scale,

$Fig.7$ in the ratio given by λ. The axiom asserts that the possibility of

any transformation implies the possibility of every similar transformation. While this axiom can be taken to have the character of a special
assumption, nevertheless any violation of it can be given an explanation
which will enable its plausibility to be preserved. It is common in
economics to make assertions on condition of "other things being equal".
Here the opposite has to be done. If, with other things equal - and
with no control over them they might be - and the transformation from
x to y is possible, but not that from $x\lambda$ to $y\lambda$ as would then be required
by the axiom, it can be argued that not all the essential coordinates
in the transformation have been made explicit, that there are other
things involved, which could be denoted by z, and are among those other
things which remain equal. The possible transformation from x to y is
thus more explicitly (x,z) to (y,z), and the other transformation is
$(x\lambda,z)$ to $(y\lambda,z)$, which only has the aspect of similarity in regard to
some coordinates but not all, so the axiom does not really apply, and
is not really violated. Thus, since z could be the rest of the world,
there is always room for a defense of the plausibility of the axiom.
There is a parallel between an axiom such as this and Newton's 'First
Law':"A particle persists in a state of uniform motion unless it is
compelled to change that state by the action of a force on it." Suppose
a particle is observed with acceleration, but there is no evidence of
a force acting on it according to familiar and general rules, say the
gravitational pull with some other particle. The First Law is not
then overthrown. It is said that mysterious forces are acting on the
particle.

As must be supposed with any irrefutable axiom, it is vacuous of
simple content. It is a formula in a way of looking at things, of
organizing perception, and through that it has its investment with
content. The general Newtonian way is very successful, but its
achievement is together with various principles, like the laws of
contact, of inertia and gravitation, for the determination of particular forces, without which it could have no result. Spectacularly,
Adams invented a mathematical planet to account for a residual motion
of Uranus which could not be explained by gravitation with the known
elements of the solar system, and it turned out to be a real one,
Neptune. Nothing is proved by such a success, but it is very satisfactory, and an encouragement to go on with the system. In economics,
as in other sciences, the concepts are laid down to give a way of pre-
senting the structure of the economic world, and rationalizing it as
arising from basic elements and their interconnection. Basic economic
interconnections are primitive and transparent to immediate comprehen-
sion, but their resultant together is not. Fitting ways have to be
found to state them, and to elicit the properties they have together,
especially those which depend essentially on form. It can seen that,
for instance, mechanics is more rooted in empiricism than economics.
But, taken at the same level, they are on the same footing. At the
most basic level, which is the level of pure economic theory, where
empiricism enters it is an empiricism of form and structure, seen or
accepted as transparent in experience.

A relation $T \subseteq \Omega \times \Omega$, considered as a set of vectors (x,y) with partitioned components $x, y \epsilon \Omega$, is a convex cone provided it satisfies the conditions

$$(x,y) \epsilon T \ \& \ \lambda \geq o \ \Rightarrow \ (x,y)\lambda \epsilon T,$$

$$(x_o, y_o), \ (x_1, y_1) \epsilon T \ \Rightarrow \ (x_o, y_o) + (x_1, y_1) \epsilon T.$$

But

$$(x,y)\lambda = (x\lambda, y\lambda), \quad (x_o, y_o) + (x_1, y_1) = (x_o + x_1, y_o + y_1).$$

Hence, with xTy denoted $(x,y) \epsilon T$, these conditions are

$$xTy \ \& \ \lambda \geq o \ \Rightarrow \ (x\lambda)T(y\lambda),$$

$$x_o T y_o \ \& \ x_1 T y_1 \ \Rightarrow \ (x_o + x_1)T(y_o + y_1),$$

that is, the Axiom of Similarity, and then the Axiom of Additivity which appeared as a consequence of the Axioms of Procession and Separation. Accordingly, the Axioms of Procession, Separation and Similarity together imply that T is a convex cone. This is the form of transformation-possibility relation which is basic in the von Neumann model.

4. Axiom of Annihilation: xTo

The stock of zero quantities of all goods, that is of nothing at all, is $o \epsilon \Omega$. The axiom asserts that any possession $x \epsilon \Omega$ can be annihilated, that is transformed into the null-possession o. Since economic possession is the first basis for economic action, the annihilation of possession of an economic agent means the annihilation of that agent as such. Thus for a firm it means final dissolution, or, for a consumer-labourer-proprietor individual, it is nothing short of suicide, unless the possibility is allowed for an enslaving subordination to some other agent. A particular consequence of this axiom is that oTo, that nothing can be transformed into nothing. But, by the Axiom of Separation,

$$oTo \text{ \& } x \geq o \Rightarrow (o+x)T(o+x).$$

Thus the Axioms of Separation and Annihilation in conjunction give xTx, for all $x \epsilon \Omega$, that is, T is a reflexive relation. This condition can be called the Axiom of Identity. It means any stock, with nothing done to it, is what it is. It must be considered one of the inseparable properties of a transformation-possibility relation, and were it not deducible from other axioms which are going to be assumed, it would have to be assumed independently. Now further, the Axiom of Separation gives

$$zTo \text{ \& } y \geq o \Rightarrow (z+y)T(y+o),$$

and hence, with the Axiom of Annihilation, it gives

$$x \geq y \Rightarrow xTy.$$

If $x \geq y$, where $x,y \epsilon \Omega$, y can be called a part of x, and $z = x-y \epsilon \Omega$ is then the residual, or complementary part. The condition asserts that any stock can be reduced to any part of it, in which case, what is the same, the residual part can be said to have been disposed of. It will be called the Axiom of Reduction. Together with the Axiom of Procession it gives

$$x \geq a \text{ \& } aTb \text{ \& } b \geq y \Rightarrow xTa \text{ \& } aTb \text{ \& } bTy \Rightarrow xTy,$$

and hence

$$aTb \text{ . } \Rightarrow \text{ . } x \geq a \text{ \& } b \geq y \Rightarrow xTy.$$

This condition can be called the Axiom of Containment. It asserts that x can be transformed into y if x contains a part which can be transformed into a stock containing y as a part. It gives, in particular,

$$xTx \; . \; \Rightarrow \; . \; x \geq x \; \& \; x \geq y \; \Rightarrow \; xTy \; .$$

The Axiom of Reduction is thus recovered from the conjunction of the Axioms of Identity and Containment. (The axioms of reduction and containment are both sometimes called axioms of free disposal.[11] Here since they have to be distinguished, they are called differently.)

5. Axiom of Economy: $x \leq y \; \Rightarrow \; x\overline{T}y$ [11a]

If $x \leq y$, which is to say all $x_r \leq y_r$ and some $x_r < y_r$, that is, no quantities of goods in y are less than in x and some are greater, then, in transformation from x to y, there would be a gain in some goods and no loss in any. The axiom asserts that such a transformation is impossible.[12] In other words, possible transformations suffer from the restriction that, with them, there can be no gain without a loss, no output without an input. Another way of stating the axiom is

$$x \leq y \; \& \; xTy \; \Rightarrow \; x=y.$$

A particular consequence is

$$oTx \; \Rightarrow \; x=o,$$

that is, only nothing can result from a possible transformation of nothing. Suppose now the contrary of the axiom, that is, with $z = y-x$, that $xT(x+z)$ for some $x \in \Omega$ and $z \geq o$, so every $z_r \geq o$ and some $z_r > o$. Then, by the Axiom of Separation, $(x+z)T((x+z)+z)$, that is $(x+z)T(x+2z)$. Then, by the Axiom of Succession, $xT(x+z)$ with $(x+z)T(x+2z)$ gives $xT(x+2z)$. Generally, $xT(x+mz)$ for $m = 1,2,\ldots$, as appears by induction. For, assuming $xT(x+mz)$ as inductive hypothesis $H(m)$, the Axiom of

Separation gives $(x+z)T((x+mz)+z)$, that is $(x+z)T(x+(m+1)z)$, and, by the Axiom of Procession, this together with the hypothesis $xT(x+z)$ gives $xT(x+(m+1)z)$, that is $H(m+1)$. Thus $H(m) \Rightarrow H(m+1)$, and since $H(1)$ is true by hypothesis, the proof by induction is complete. Thus, the Axioms of Procession and Separation, and the negation of the Axiom of Economy, in conjunction, imply that, for some $x \epsilon \Omega$ and $z \geq o$, $xT(x+mz)$ for all $m = 1,2,...$ In this case, starting with possession of x, and by possible transformations, any good r for which $z_r > o$ can be possessed in arbitrarily large quantity, without sacrifice. Such goods therefore are free, being without cost in terms of necessary sacrifice for their gain; and they are not scarce, being available at will in limitless amounts. They are therefore without the essential characteristics of economic goods. It follows that, if the Axioms of Procession and Separation are to be assumed, and if economic goods alone are to be allowed to enter into the description of a stock, then the Axiom of Economy is required.

6. Axiom of Continuity:[13] T is closed in $\Omega \times \Omega$

The points of $\Omega \times \Omega$ are (x,y) where $x,y \epsilon \Omega$, and are identified with vectors thus given in partitioned form. Convergence of a sequence of vectors is defined by the simultaneous convergence of the associated sequences of elements. A set of vectors is closed provided that, for any sequence of points in it which converges to a limit, if each point of the sequence belongs to the set then so does the limit. In particular, $\Omega \times \Omega$ is closed. Now T is defined as a set in $\Omega \times \Omega$, and the axiom asserts that it is a closed set. Thus, given $(x_r,y_r) \epsilon T$, $r = 1,2,...$,

such that $(x_r, y_r) \rightarrow (x,y)$ $(r \rightarrow \infty)$, if $(x_r, y_r) \in T$, $r = 1, 2, \ldots$
then $(x,y) \in T$. Since xTy denotes $(x,y) \in T$, another way of stating
this is that, given x_r, $y_r \in \Omega$, $r = 1, 2, \ldots$ such that $x_r \rightarrow x$,
$y_r \rightarrow y$ $(r \rightarrow \infty)$, if $x_r T y_r$ for $r = 1, 2, \ldots$ then xTy. There is a simple
practical basis for this axiom. A stock is defined by amounts of
goods which compose it. Any amount of a good is recorded to within
some perhaps small, but nevertheless definite quantity, say some fraction of a unit. In other words, quantities are rounded to a multiple
of some fraction of a unit part, or quantum of measurment, usually "the
last decimal place" in the record of most practical measurement. This
means that there ia a practical identification between any stocks in
which amounts of goods differ by sufficiently small quantities; and
this means that, if $(x_r, y_r) \rightarrow (x,y)$ $(r \rightarrow \infty)$, then, in terms of practical
accounts, this must signify that for some N and for all $r > N$, $(x_r, y_r) =$
(x,y). But now there is no practical distinction between the statements
$(x_r, y_r) \in T$ for $r > N$, and $(x,y) \in T$. This shows what might be called
the phenomenalogical meaning, and essential content, of the axiom. So
it appears that there can be no real substance to a denial of the axiom,
just as there is only this little substance to its assertion. Assumption
of the axiom is a mathematical way of accomodating the practical reality
of quanta of measurement without being specific about them.

The equivalence has already been remarked between the Axioms of
Similarity and Rational Similarity, on condition of the Axiom of Continuity, and it will now be shown. Since obviously one implies the
other as a special case, it remains to show the converse implication.
Thus, assume the Axioms of Rational Similarity and Continuity. Let

xTy and $\lambda \geq o$. It has to be shown that $(x\lambda)T(y\lambda)$. For any real
$\lambda \geq o$ there exist rationals $\rho_r \geq o$, $r = 1,2...$, such that $\rho_r \to \lambda$ $(r \to \infty)$
Let $x_r = x\rho_r$, $y_r = y\rho_r$, $r = 1,2,...$ Then $x_r \to x\lambda$, $y_r \to y\lambda$ $(r \to \infty)$.
But by the rational similarity axiom, $x_r T y_r$, $r = 1,2,...$ Now by the
continuity axiom, $(x\lambda)T(y\lambda)$.

Six axioms have been proposed, independently, each with its
separate meaning, and each with its own recommendation. They have
already been seen as necessary properties of any transformation-
possibility relation T having the form $T = I_U$, where

$$I_U = \{(x,y) : u'x \geq u'y \quad \text{for all } u \epsilon U\},$$

and U is a non-negative, closed, convex cone. It is about to appear
that together they are entirely characteristic of this form. That is,
not only does a relation with this form necessarily have these
properties, but also if it has these properties then it must necessarily
have this form.

5: Characterization of Normal Structure

A normal transformation-possibility relation is defined to be any relation $T \subseteq \Omega \times \Omega$ which satisfies the following axioms.

Axiom 1: $xTyTz \Rightarrow xTz$

Axiom 2: xTy & $z \geq o \Rightarrow (x+z)T(y+z)$

Axiom 3: xTy & $\lambda \geq o \Rightarrow (x\lambda)T(y\lambda)$

Axiom 5: xTo

Axiom 6: T is a closed set in $\Omega \times \Omega$

Consequences are as follows, where xT, $Ty \subseteq \Omega$, such that

$$xTy \Leftrightarrow x\epsilon Ty \Leftrightarrow y\epsilon xT,$$

define the forward and backward transformation sets for any x, $y\epsilon\Omega$.

Proposition 1: xTx

That is, T is a reflexive relation. By Ax. 4, oTo, and by Ax. 2, $oTo \Rightarrow (x+o)T(x+o)$, whence the proposition.

Thus T, being reflexive, by this proposition, and transitive, by Ax. 1, is an order.

Corollary 1.1: $xTy \Rightarrow x,y \epsilon xT\cap Ty$

Proposition 2: $xT\cap Ty \neq 0 \Rightarrow xTy$

For, by Ax. 1, $z \epsilon xT\cap Ty \Rightarrow xTz$ & $zTy \Rightarrow xTy$.

Proposition 3: $xTy \Leftrightarrow xT \supseteq yT \Leftrightarrow Tx \subseteq Ty$

By Cor. 1.1, $y \epsilon yT$, so that $yT \subseteq xT \Rightarrow y \epsilon xT \Rightarrow xTy$. By Ax. 1, xTy .$\Rightarrow$. $yTz \Rightarrow xTz$, that is $xTy \Rightarrow yT \subseteq xT$. Similarly for Tx.

This shows $x \to xT$ is a homomorphism of the T-order of Ω into the inclusion-order of a set of subsets of Ω.

The reversible transformation-possibility relation R which de-rives from T is the symmetric part $R = T \cap T'$ of T. Thus

$$xRy \Leftrightarrow x(T \cap T')y \Leftrightarrow xTy \ \& \ xT'y \Leftrightarrow xTy \ \& \ yTx.$$

Corollary 3.1: $xRy \Leftrightarrow xT = yT \Leftrightarrow Tx = Ty$

Proposition 4:   R is an equivalence

From the form of its definition, R is symmetric. It is reflexive and transitive, from the same properties for T given by Ax. 1 and Prop. 1. Hence R is an equivalence; and this is also immediately ob-vious from Cor. 3.1.

Now Ω is partitioned into R-classes, and T reduces to a simple order T/R of these classes.

Proposition 5: $(xT)+y \subseteq (x+y)T$, $(Tx)+y \subseteq T(x+y)$

In this notation, $(xT)+y = \{a+y \ : \ a \epsilon xT\} = \{a+y \ : \ xTa\}$. So if $z \ \epsilon \ (xT)+y$ then $z-y \ \epsilon \ xT$, that is $xT(z-y)$. But now, by Ax. 2, $(x+y)T((z-y)+y)$, that is $z \ \epsilon \ (x+y)T$. Hence $(xT)+y \subseteq (x+y)T$; and simi-larly for Tx.

Proposition 6: $(xT)\lambda = (x\lambda)T$, $(Tx)\lambda = T(x\lambda)$ $(\lambda > o)$

In this notation, $(xT)\lambda = \{a\lambda \ : \ a \epsilon xT\} = \{a\lambda : xTa\}$. So if $y \ \epsilon \ (x\lambda)T$, that is $(x\lambda)Ty$, then, by Ax. 3, $xT(y\lambda^{-1})$, that is $y\lambda^{-1} \ \epsilon \ xT$, that is $y \ \epsilon \ (xT)\lambda$. Hence $(x\lambda)T \subseteq (xT)\lambda$. Now replace x here by $x\lambda$ and λ by λ^{-1}. Then $xT = ((x\lambda)\lambda^{-1})T \subseteq ((x\lambda)T)\lambda^{-1}$. Accordingly, $(x\lambda)T = (xT)\lambda$; and similarly for Tx.

Proposition 7:   xT, Tx are convex

If $y, z \epsilon xT$, that is xTy, xTz, and if $\lambda, \mu \geq o$, then by Ax. 3, $(x\lambda)T(y\lambda)$, $(x\mu)T(z\mu)$. Now, by Ax. 2, $(x\lambda+x\mu)T(y\lambda+x\mu)$ and $(x\mu+y\lambda)T(z\mu+y\lambda)$, and hence, by Ax. 1, $(x(\lambda+\mu))T(y\lambda+z\mu)$. So if, moreover, $\lambda+\mu = 1$, then

$xT(y\lambda+z\mu)$, that is $y\lambda+z\mu \in xT$. Thus $y,z \in xT$ & $\lambda,\mu \geq o$ & $\lambda+\mu=1$ $\Rightarrow$

$y\lambda+z\mu \in xT$. With $< y,z > = \{y\lambda+z\mu : \lambda,\mu \geq o, \lambda+\mu=1\}$, this states

$y,z \in xT \Rightarrow < y,z > \in xT$. Thus xT is convex; similarly Tx is convex.

Corollary 7.1: $xT \cap Ty$ is convex

For the intersection of convex sets is a convex set.

Proposition 8: $xTy \Rightarrow < x,y > \subseteq xT \cap Ty$.

By Cors. 1.1 and 7.1, $xTy \Rightarrow x,y \in xT \cap Ty \Rightarrow < x,y > \subseteq xT \cap Ty$.

Proposition 9: xT, Tx are closed sets

Let $y_r \in xT$ $(r = 1,2,\ldots)$ and $y_r \to y$ $(r \to \infty)$. It has to be shown

that then $y \in xT$. Now $(x,y_r) \in T$ and $(x,y_r) \to (x,y)$. But, by Ax. 6, T

is closed. It follows that $(x,y) \in T$, that is $y \in xT$. Thus xT is closed;

and similarly Tx is closed.

Proposition 10: $x \geq y \Rightarrow xTy$

If $y \geq x$, then $z = x-y \geq o$ and hence, by Ax. 5, zTo. Now, by

Ax. 2, $(z+y)T(o+y)$, that is xTy.

The relation $D \subseteq \Omega x\Omega$ defined by $xDy \Leftrightarrow x \geq y$ & $x,y \in \Omega$, can be

viewed both as a relation of domination and of disposal. Thus, if xDy,

then y is dominated as a part of x, and also is obtained from x by

disposal of the residual part $x-y$. Another statement of Prop. 10 is

$D \subseteq T$, or again, $xD \subseteq xT$. If $x > o$, the interior of xD is non-empty.

Hence also the interior of xT is non-empty, since it contains it. Now

it follows from Prop. 3 that $y \in xT \Rightarrow yD \subseteq xT$. Therefore, since $y \in yD$,

it follows that

$$xT = \cup\{yD : y \in xT\}.$$

That is, the form of xT is such that it is identical with the set of

all points dominated by any of its points. Thus if xT contains any

point y > o, its interior must be non-empty, since it then contains
yD which has its interior non-empty. But in any case, there is the
following.

Corollary 10.1: If x > o, then xT has non-empty interior.

Proposition 11: xT is bounded

By Prop. 7, xT is convex. Since Ω is convex, x+Ω is convex. By
Ax. 5, x $\leq$ y & xTy $\Rightarrow$ x=y, that is, xT$\cap$(x+Ω) = {x}. Thus xT and x+Ω
are two convex sets whose intersection is a single point x. It follows,
from the theorem on the separating hyperplane, that there exists a
hyperplane through x separating their interiors. This means there
exists a vector u such that

$$y \epsilon xT \Rightarrow u'y \leq 1, \quad y > x \Rightarrow u'y > 1, \quad u'x = 1.$$

But y > x $\Rightarrow$ u'y > u'x implies u > o, and this implies the set
{y : u'y $\leq$ 1, y $\geq$ 0}, containing xT, is bounded. Hence xT is bounded.

In xTy, y is maximal if z $\geq$ y & xTz $\Rightarrow$ z=y; and x is minimal if
z $\leq$ x & zTy $\Rightarrow$ z=x.

Proposition 12: If y is maximal in xTy then x is minimal.

If x is not minimal then zTy for some z $\leq$ x. Now x-z $\geq$ o, and
hence, by Ax. 2, (z+(x-z))T(y+(x-z)), that is xT(y+(x-z)), where
y+(x-z) $\geq$ y, that is y is not maximal.

The efficiency frontier of xT is described by the maximal points
in xTy. The efficient transformation-possibililily relation which
derives from T is the relation E such that xEy provided xTy and y is
maximal; so E $\subseteq$ T, and xE, lying in xT, is identical with the frontier
of xT. Since yϵxE & z $\geq$ y $\Rightarrow$ zϵx$\overline{T}$, points y on the frontier in xT
are arbitrarily close to points z of the complement x$\overline{T}$ of xT in Ω, so
they are part of the boundary of xT. Points which are on the boundary

of xT and also on the boundary of Ω describe the perifery of xT.
Boundary points of xT which are both on the frontier and the
perifery describe the frontier-perifery of xT. By Cor. 10.1, if
x > o, the interior of xT is non-empty. The frontier xE is then a
convex surface zone, bounded by the frontier-perifery, which separates
the interior of xT from the complement $\overline{xT}$. Since xT is bounded, every
point z of xT is dominated by some point y $\geq$ z of xE, so

$$xT = \cup\{yD : xEy\}.$$

In this sense, xE is a dominating subset of xT. Also it is a minimal
dominating subset, since, by definition, the points of xE are undominated in xT.

Proposition 13: If y lies on the frontier of xT, then so does every
point of the segment < x,y >.

Another statment is xEy $\Rightarrow$ xE < x,y >. Let y be on the frontier
of xT, that is yϵxT, but $\epsilon \geq$ o $\Rightarrow$ y+ϵ ϵ $\overline{xT}$. Then, by Prop. 8,
< x,y > ϵ xT$\cap$Ty, but, by Prop. 2, xT$\cap$T(y+ϵ) = 0. Now, by Prop. 5,
(Ty)+ϵ $\subseteq$ T(y+ϵ). Therefore xT$\cap$((Ty) + ϵ) = 0, that is (Ty)+ϵ $\subseteq$ $\overline{xT}$,
and, since < x,y > + ϵ $\subseteq$ Ty+ϵ, this implies < x,y > + ϵ $\subseteq$ $\overline{xT}$. Accordingly, < x,y > $\subseteq$ xT, but $\epsilon \geq$ o $\Rightarrow$ < x,y > + ϵ $\subseteq$ $\overline{xT}$, which was to be
proved.

Proposition 14: For x > o, xT is the intersection in Ω of a cone
with vertex x.

Fig. 8)   Consider the intersection ρ in Ω of any ray with vertex v > o.
It has to be shown that, excepting x, ρ must lie entirely inside or
outside xT. If possible, suppose the contrary, for some ρ. Since
xT is compact, by Props. 9 and 11, and convex, by Prop. 7, its

intersection with ρ will be a compact segment $< x,y > = ρ ∩ xT$. By
hypothesis, $y \neq x$. Now $y \geq x$ is impossible, by Ax. 5, because xTy.
It follows that ρ meets the boundary of Ω in some point e. By hypothesis, $e \neq y$, for otherwise $ρ ⊆ xT$. Thus $ρ = < x,e > ⊂ < x,y >$.
Now $e_\epsilon \overline{xT}$, so that, by Ax. 5, $e \neq o$. Also $o_\epsilon xT$, by Ax. 4; moreover,
since $x > o$, o is not on the frontier of xT. Therefore the frontier
separates o, e and cuts $< o,e >$ in a point f on its perifery. Thus
$< o,e > ∩ xT = < o,f > ⊂ < o,e >$. Now the four distinct points
o,x,y,f form a regular coplanar quadrangle, since no three are collinear,
and a pair of opposite sides, o,f and x,y, intersect, in e. Therefore
the diagonals o,y and x,f intersect in a further point, say z. Now z
in the interior of $< o,y >$ implies $z \leq y$. Therefore, since y is in
xT and properly dominates z, z cannot be on the frontier of xT. But
also $z_\epsilon < x,f >$, and f is on the frontier. By Prop. 13, this means z
must be on the frontier. Hence there is a contradiction, and the
original supposition about ρ is impossible, so the proposition is
proved.

Any $z = y-x$ such that xTy defines a forward transformation displacement for T at x. Similarly $y-x$ such that yTx defines a backward
displacement. Thus $\nabla_x = xT-x$, $\Delta_x = Tx-x$ are the sets of forward and
backward transformation displacements for T at x. The ray $\vec{z} =$
$\{zλ : λ \geq o\}$ through any transformation displacement z defines a transformation direction. Any collection of rays forms a cone. The cone
$\vec{S}$ through any set S is formed by the set of rays through its elements,
so

$$\vec{S} = ∪\{\vec{z} : z_\epsilon S\} = \{zλ : z_\epsilon S, λ \geq o\}.$$

If S is convex, or closed, so is $\bar{S}$. Obviously, for any $\lambda > o$, the cones through S and $S\lambda = \{z\lambda : z\epsilon S\}$ are the same: $(\vec{S\lambda}) = \vec{S}$. Also, if $R \subseteq S$, then $\vec{R} \subseteq \vec{S}$. The cones of forward and backward transformation directions for T at x are given by $\vec{\nabla}_x$, $\vec{\Delta}_x$. Prop. 14 gives

$$xT = \Omega\cap(x+\vec{\nabla}_x).$$

The following lemmas prepare for the final result.

Lemma 1: $\nabla_{x\lambda} = \nabla_x\lambda$

For $\nabla_{x\lambda} = (x\lambda)T-x\lambda = (xT)\lambda-x\lambda = (xT-x)\lambda$, by Prop. 6.

Lemma 2: $x \leq y \Rightarrow \nabla_x \subseteq \nabla_y$

For, by Prop. 5, $y-x \geq o$ implies $(xT) + (y-x) \subseteq (x+(y-x))T$, that is $(xT)+(y-x) \subseteq yT$, that is $xT-x \subseteq yT-y$.

Lemma 3: $x, y > o \Rightarrow \vec{\nabla}_x = \vec{\nabla}_y$

If $x, y > o$ then there exist $\lambda, \mu > o$ such that $y\lambda < x < y\mu$. Then, by Lemma 2,

$$\nabla_{y\lambda} \subseteq \nabla_x \subseteq \nabla_{y\mu} \, ,$$

so that, by Lemma 1,

$$\nabla_y\lambda \subseteq \nabla_x \subseteq \nabla_y\mu \, .$$

Now $(\vec{\nabla_y\lambda}) = \vec{\nabla}_y$, and so forth. Hence, taking cones,

$$\vec{\nabla}_y \subseteq \vec{\nabla}_x \subseteq \vec{\nabla}_y \, ,$$

as required.

It follows from Lemma 3 that there exists a cone V, determined by T, such that

$$x > o \Rightarrow \vec{\nabla}_x = V.$$

Since xT is closed and convex, by Props. 7 and 9, V is closed and convex. Now Prop. 14 gives

$$x > o \quad \Rightarrow \quad xT = \Omega \cap (x+V),$$

that is,

$$x > o \quad .\Rightarrow. \quad xTy \quad \Leftrightarrow \quad y-x \epsilon V \ \& \ y \epsilon \Omega.$$

Since, by Ax. 6, T is closed in $\Omega x \Omega$, this gives

$$xTy \quad \Leftrightarrow \quad y-x \epsilon V \ \& \ x,y \epsilon \Omega.$$

That is, in a previous notation (section 2), $T = E_V$. By Prop. 11, xT is bounded, and this implies the existence of the dual cone U of V, given by

$$U = \{u : z \epsilon V \Rightarrow u'z \leq o\}.$$

Then, by the duality theorem for convex cones,

$$V = \{z : u \epsilon U \Rightarrow u'z \leq o\},$$

that is,

$$z \epsilon V \quad .\Leftrightarrow. \quad u \epsilon U \Rightarrow u'z \leq o.$$

Accordingly,

$$xTy \quad .\Leftrightarrow. \quad y-x \epsilon V \ \& \ x,y \epsilon \Omega$$

$$.\Leftrightarrow. \quad u \epsilon U \Rightarrow u'(y-x) \leq o \ \& \ x,y \epsilon \Omega$$

$$.\Leftrightarrow. \quad u \epsilon U \Rightarrow u'y \leq u'x \ \& \ x,y \epsilon \Omega.$$

Thus

$$xTy \quad .\Leftrightarrow. \quad u'y \leq u'x \ \text{for all} \ u \epsilon U \ \text{and} \ x,y \epsilon \Omega$$

That is, in a previous notation (section 2), $T = I_U$. Now, by Prop. 10, $x \geq y = xTy$. Therefore $u \epsilon U$ implies $u'y \leq u'x$ for all $y \leq x$; and this implies $u \geq o$ for all $u \epsilon U$, that is $u \subseteq \Omega$. In order that the condition $oTx \Rightarrow x=o$, required by Ax. 5, be satisfied, it is necessary that u'x $\leq$ o for all uϵU $\Rightarrow$ x=o,

and this requires U to cut the interior of Ω. It would be enough
is U were regular, that is of full diemnsion, or with non-empty
interior. But all that is asked is that U does not lie in a face
of Ω. Now the following has appeared.

THEOREM: A necessary and sufficient condition for a transformation-
possibility relation T to be normal is that there exists a closed
convex cone U in and cutting the interior of Ω such that xTy if and
only if x,y$\epsilon\Omega$ and u'x $\geq$ u'y for all uϵU.

In case U is a finitely generated covex cone, say with k generators
$u_1,\ldots,u_k \geq$ o, so that

$$U = \{\Sigma u_r \lambda_r : \lambda_r \geq o\} = < u_1, \overrightarrow{\ldots,} u_k > ,$$

then T, which will then also be said to be finitely generated, is
such that

$$xTy \leftrightarrow u_r'x \geq u_r'y \quad (r = 1,\ldots,k) \; \& \; x,y \geq o.$$

Thus T is defined by a system of simultaneous homogeneous linear inequalities in non-negative variables. This is an analytically con-
venient form to assume for a transformation-possibility relation;
moreover, any normal relation can be uniformly approximated in any
compact region by relations of this form. A closer approximation is
obtained from any given approximation just by a suitable enlargement
of the set of generators. However, the finitely generated form has
a universal merit of its own, if it is granted that any data which
could be available for the empirical construction of the relation
must necessarily be finite.

With

$$U = (u_1, \ldots, u_k)$$

now denoting the nxk-matrix formed with the k generating n-vectors
as columns, a relation T such that

$$xTy \quad \Leftrightarrow \quad U'x \geq U'y \ \& \ x,y \geq o,$$

for some matrix $U \geq 0$ with no row or column null gives the matrix
statement of the general form of a finitely-generated normal transformation-possibility relation.

It is possible to manifest the basic features and structures of
economic theory, of production, consumption, capital, labour, exchange
and equilibrium, by appropriate assumption of relations of this general
form, always keeping to finite, constructive algebraical statements
and proceeding entirely within the framework of theory of linear pro-
grams. In this way, without loss and with a definite advantage, the
methods of the differential calculus are rendered inessential, while
at the same time they are kept with their proper bearing.

A basic illustration is provided by derivation of features of production theory, in particular the classical production function, which
will appear with a natural generalization, stated with a very workable
analytic form. It will be seen that the present method brings a new
foundation, through the axiomatic analysis, and also a certain enlargement and synthesis to standard discussion, and is quite simple.

6: Input-Output

Consider stocks of some n+m goods. Let the goods be separated into two classes, n to be inputs and m to be outputs. For simplicity, the classes have been supposed not to overlap, though this is not essential. Correspondingly, any stock $z \epsilon \Omega^{n+m}$ of the n+m goods will have representation $z = (x,y)$ where $x \epsilon \Omega^n$ is a stock of goods of the input class and $y \epsilon \Omega^m$ is of the output class. Let $T \subseteq \Omega^{n+m} \times \Omega^{n+m}$ be any transformation-possibility relation for the n+m goods. Consider the possible transformations between an initial stock of the form (x,o) and a final one of the form (o,y). That is, initially only input goods are held, and in the transformation they are used up, and a stock of goods of the output class is produced instead. Thus, an input-output relation $IO \subseteq \Omega^n \times \Omega^m$, induced by T with this partition between input and output goods, is defined by

$$xIOy \Leftrightarrow (x,o)T(o,y).$$

The output-possibility set for a given input $x \epsilon \Omega^n$ is $xIO \subseteq \Omega^m$, where $y \epsilon xIO \Leftrightarrow xIOy$. Similarly, the input-possibility set for a given output y is IOy. The relation $I\overset{.}{O} \subseteq IO$, by which an output y is efficiently related to a given input x, is defined by

$$xI\overset{.}{O}y \;.\Leftrightarrow.\; z \geq y \;\&\; xIOz \Rightarrow z=y.$$

Similarly, the relation $\overset{.}{I}O$, by which an input is efficiently related to a given output, is defined by

$$x\overset{.}{I}Oy \;.\Leftrightarrow.\; z \leq x \;\&\; zIOy \Rightarrow z=x.$$

The efficient-output-possibility set for a given input x is $xI\overset{.}{O}$; similarly, $\overset{.}{I}Oy$ is presented. The relation an input and an output have in case they are each efficiently related to the other can be denoted $\overset{..}{\text{IO}}$. These are the main general concepts associated with an input-output relation.

If T is a normal transformation-possibility relation, then the relation IO induced by it will be called a __normal input-output relation__. The structure of such a relation now has to be examined. With T normal, there exists a matrix W with semi-positive rows and columns, such that

$$XTY \iff WX \geq WY \; \& \; X,Y \geq 0.$$

With partitions at the nth row, let

$$X = \begin{pmatrix} x_o \\ x_1 \end{pmatrix}, \quad Y = \begin{pmatrix} y_o \\ y_1 \end{pmatrix},$$

and with partition at the nth column let $W = (U,V)$. Then,

$$XTY \iff Ux_o + Vx_1 \geq Uy_o + Vy_1, \; x_o, y_o \geq o, \; x_1, y_1 \geq o.$$

Thus, with $x_o = x$, $x_1 = o$ and $y_o = o$, $y_1 = y$, it appears that

$$xIOy \iff Ux \geq Vy \; \& \; x \geq o \; \& \; y \geq o,$$

where U,V have semipositive columns, and (U,V) has semipositive rows. That is, with $A_{(r)}$, A_s denoting row r and column s of any matrix A,

$$(U_{(r)}, V_{(r)}) \geq o, \; U_s \geq o, \; V_t \geq o \quad (r = 1,\ldots,k; \; s = 1,\ldots,n; \; t = 1,\ldots,m).$$

This states the general form of a normal input-output relation.

The output-possibility set $Y = F(x)$, determined as a function $F(x) = xIO$ of input x, is the appropriate extension for several products of the standard concept of a production function for a single product.[14] Its structure will be examined on the assumption of normality, and it will be seen then to give a direct generalization of the classical production function for the case of constant returns to scale, such a function being taken to have the form of a non-decreasing

homogeneous concave function. Account of the case of non-decreasing

returns to scale, and correspondingly of a general non-decreasing

concave production function, is also readily given, within the frame-

work of the case of constant returns. It can be remarked that the

common general assumption of differentiability, or worse, continuous

differentiability, for a production function is objectionable, in that,

with it, it is impossible to have representation of some most essential

features of production, such as complementarities among inputs. The

methods here do not suffer from that objection. Moreover they are more

readily available for empirical and computational realization.

Let $x \geq o$ denote a fixed input. The associated output-possibility

set is

$$xIO = \{y : U_{(r}x \geq V_{(r}y \ (r = 1,\ldots,k), \ y \geq o\}.$$

Since $xIO \subseteq \{o\}$, it is not empty. Since $V_{s)} \geq o \ (s = 1,\ldots,m)$, it is

bounded. Thus it is a bounded convex polyhedral region, any face of

which is a convex polyhedron lying either in one of the coordinate

hyperplanes of Ω^m, or in one of the hyperplanes $\{y : U_{(r}x = V_{(r}y\}$.

Since $U_{(r}x \geq o, V_{(r} \geq o$, this establishes it with the form of a classi-

cal production-possibility set. Two possible special cases are worth

noting. One is $xIO = \{o\}$. This can arise, for instance, if any

$U_{(r}x = o$ while $V_{(r} > o$. No output is possible with such an input x.

Thus, to have some output, it might not be enough just to have some in-

put. The input might also have to have the right composition - thus,

some machines, without some labour, are useless. Another possible

case, which includes the foregoing, is $xIO = \{y : o \leq y \leq b\}$, for some $b \geq 0$. In this case there is just one efficient output, $y=b$. It can be said in this case that there is no substitutability in the output, with that input, no more of one good can be produced by sacrificing a quantity of some other. In any other case, there is substitutability.

Efficient outputs $\dot{y}\epsilon xIO \subseteq xIO$ can be characterized by relation with output prices $p > 0$. Thus, xIO being bounded, let

$$pM_x\dot{y} \quad \Leftrightarrow \quad p'\dot{y} = \max\{p'y : xIOy\}$$

define the relation M_x which prices p have to output $\dot{y}$ which is optimal, in that they give maximum return $p'\dot{y}$ at those prices, x being input. Then

$$\dot{y}\epsilon xIO \quad \Leftrightarrow \quad pM_x y \text{ for some } p > 0,$$

that is, efficiency is equivalent to optimality at some positive prices. Efficient points y in xIO all obviously must satisfy $U_{(r}x{=}V_{(r}y$ for some r. However, not all such points are efficient. The efficient points are part of the forward boundary of xIO, and describe a connected zone in a convex polyhedral surface, which is "convex away from the origin". They form a dominant set in xIO; each is undominated, and every other point in xIO is dominated by at least one of them.

Now consider the case $m=1$, in which there is just one output good. For given input x, efficient output requires the quantity to be a maximum $F(x)$, where

$$F(x) = \max\{y : U_{(r}x \geq V_r y \quad (r = 1,\ldots,k), y \geq 0\}$$

$$= \max\{y : 0 \leq y \leq \frac{1}{V_r} U_{(r}x ; V_r > 0\}$$

$$= \min\{\frac{1}{V_r} U_{(r}x : V_r > 0\},$$

where $V_r > o$ for some r, since V is a semipositive column vector.
From $U_{(r} \geqq o$ it follows that $\frac{1}{V_r} U_{(r}x$ is a non-decreasing homogeneous
linear function of x. Now $F(x)$, being expressed as the minimum of
a set of such functions, appears as a non-decreasing homogeneous con-
cave function; so $F(x\lambda) = F(x)\lambda$ $(\lambda \geq o)$, which shows the condition of
constant returns to scale. This set of linear functions being finite,
$F(x)$ is a polyhedral function of that class. Without the finiteness
restriction, it would be a general function of that class. Thus, in
the framework of the proposed axioms, and from a more general concept,
the classical production function is derived for the case of constant
returns. Now suppose inputs to be partitioned as $x = (a,z)$, into a
constant component a, and a variable component z. Let $\Phi(z) = F(a,z)$.
Then $\Phi(z)$ is a non-decreasing concave function. It thus has the form
of a classical production function, for the general case without constant returns. Since $F(x\lambda)$ is then a concave function of $\lambda \geq o$, it
shows non-increasing returns to scale. The classical production function carrying with it the principle of non-increasing returns to
scale (usually formulated as a principle of decreasing returns to
scale), has thus been established on the basis of the axioms which
have been proposed.

Returning now to the normal input-output relation in general
form, consider the input-possibility set for a given fixed output
$y \geq o$. It is

$$IOy = \{x: U_{(r}x \geqq V_{(r}y \quad (r = 1,\ldots,k); \ y \geqq o\}.$$

Since $V_{s)} \geq o$ for all s, $y \geq o$ implies $V_{(r}y > o$ for some r. Also
$U_{(r} \geqq o$ for all r. It follows that IOy does not contain o. Since

the columns $U_{s)}$ of U are semipositive, IOy is unbounded. It appears

as a convex polyhedral region is bounded away from the origin by

hyperplanes $\{x:U_{(r}x = V_{(r}y\}$. Thus it appears as a region bounded from

below by a surface which has the form of a classical utility indif-

ference surface, which is "convex towards the origin". Two possible

special cases are worth noting. In one case, IOy is empty, that is

contains no finite points. This means y is impossible to produce, what-

ever the input. This case can arise, for instance, if $U_{(r} = o$ while

$V_{(r}y > o$. The other case to be noted is IOy = $\{x:x \geq a\}$, for some

$a \geq o$. This means that, for the output y, there is just one efficient

input x=a, or that there is no substitutability in input. It is im-

possible to offset a decrease in the quantity of one good in the input

a by an increase sin some others, and still produce y. In any other

case, there would be substitutability for efficient input. Efficient

inputs can be characterized as possible inputs which, for some positive

input-prices, give minimum cost.

In this general discussion, it has been supposed that "all" in-

puts and outputs are explicit, and variable. But, in any application,

"all" may have to include certain inputs and outputs which do not have

explicit recognition, which may be conditioned as part of nature, or

plant, or subsistence, or any basic resource or requirement which is

$Fig.10)$ taken for granted. Therefore it is fitting to consider the form of a

normal input-output relation which would follow on certain of the var-

iables being fixed. It is

$$xIOy \leftrightarrow Ux+A \geq Vy+B \ \& \ x \geq o \ \& \ y \geq o,$$

where now just $U_{s)} \geq o$, $V_{t)} \geq o$, and $A \geq o$, $B \geq o$, and x, y denote just the variable inputs and outputs. What is the same,

$$xIOy \iff Ux+C \geq Vy \ \& \ y \geq o,$$

where C is some vector, free of any non-negativity or other requirement, and whose constancy is on condition of the constancy of the inputs and outputs which are ignored. This is the appropriate form for when the interdependence is to be considered between "some" inputs and outputs, as is usually, or inevitably, the case. Again, in the case of just one output good, this form leads directly to the classical production function.

This subject could be elaborated further in various ways, to give account of special structures which are possible, or of modes of composition of relations to form more complex ones. But here is completed a brief and basic discussion of the input-output relations which derive from a normal transformation-possibility relation.

It is well to remark that exchange, and markets, fit with particular simplicity into the scheme. Trade, or reciprocal exchanges, redistributes goods among the traders, without adding to any total amounts. The trade-possibility relation T for k traders in n goods, is defined by

$$XTY \iff UX \geq UY \ \& \ X,Y \geq 0,$$

where

$$X = \begin{pmatrix} x_1 \\ \cdot \\ \cdot \\ \cdot \\ x_k \end{pmatrix}, \ Y = \begin{pmatrix} y_1 \\ \cdot \\ \cdot \\ \cdot \\ y_k \end{pmatrix}, \ U = (I,\ldots,I) \ ,$$

and where $x_1,\ldots,x_k,\ y_1,\ldots,y_k \in \Omega^n$, and $I = \begin{pmatrix} 1 & 0 \\ 0 & 1 \end{pmatrix}$. A market with
n goods, and prices $p > o$, gives rise to a transformation-possibility
relation T such that

$$xTy \leftrightarrow p'x \geqq p'y \ \& \ x,y \geqq o,$$

which is again normal.

$Fig.11\rangle$ A consumer has the aspect of a normal transforming agent who,
like a firm with fixed basic plant, has certain basic capacities and
equipment, who has consumer commodities as variable inputs, and whose
output consists of various goods, such as labour, and any works, pri-
vately used, or marketed along with labour to meet consumption cost,
and a further good, designated as utility, which, together with the
motive to maximize output of this good by all possible means, typifies
the consumer as an economic agent. With some inputs, work may be a
pleasure, that is utility is a complementary output. Or work and
pleasure may be substitutes; labour is output just because of its return
in money which can be expended on consumption, and, in the final balance, there will be more utility. In the usual simple view, commodities are input and utility is output. Here, the utility function
appears as an ordinary production function, and the identity of form
between a classical production and a classical utility function seems
not the accident it otherwise could seem. But further than this simple
view, complicated structures and delicate balances of possibilities
can be grasped as a whole, by means of the normal transformation-
possibility relation. In fact, all the features of an elaborately
structured economy, at least as comprehensive as the Arrow-Debreu type,

can be algebraically stated in terms of a single, though somewhat complex, normal transformation-possibility matrix, and studied from this
point of view, for such questions as general equilibrium. This will be
done elsewhere. It has been seen that in the general view of transformation, some goods involved might not be simple material goods, and
a special instance is consumers utility. While simple material goods
have a direct physical manifestation, and a corresponding observable
measure, this is not the case for utility. Its measurement, in a form
appropriate to the extent of its significance, must be based more broad-
ly on the supposition of its existence bearing on contingent observations.

Coefficients which give marginal rates of substitution, either between input goods for a fixed output, or output goods for a fixed input,
or between an input and an output good, other inputs and outputs being
fixed, are most readily treated from the point of view of linear programming theory, using duality. They are, in a certain way, associated
with corresponding ratios of elements in vectors u_r, v_r. In the simple
special case of $xIOy$ in which $x > 0$, $y > 0$, and $u_r x = v_r y$ for just one
r, say $r=t$, so $u_r x < v_r y$ for $r \neq t$, they are ratios of the form u_{ti}/u_{tj},
v_{ti}/v_{tj}, u_{ti}/v_{tj}. Such coefficients are typical concepts of the classical
theory. In fact, they are the essential contribution of that theory. But
there they proceed on assumption of differentiability, which gives these
coefficients as ratios of positive partial derivatives. A differentiable
function with positive partial derivatives is an increasing function.
An increase in any input brings about an increase in output. But in case
any inputs are complements, there would be no gain by increasing one and

not the other. H^a This brings a contradiction of monotonicity with

differentiability. The differential of the function cannot be just

a linear function of the argument differentials. It is necessary there-

fore to proceed without the general assumption of differentiability.

But this abolishes use of the straightforward methods of the differential

calculus. Instead there has to be reliance on general methods of con-

vex set theory. But convex sets can only be handled specifically by

their property of approximation by finitely generated sets. Thus in-

evitably arises the algebraical model that has now been treated. The

axiomatic analysis, however, bears on the more general normal model, which

stands above both the presently discussed finite combinatorial, or pro-

gramming method, and the classical differential or marginal method.

7: Empirical Admissibility

Consider n goods, and an agent who transforms them. Let the agent be observed in m acts of transformation, say x_t into y_t (x_t, $y_t \geq 0$; $t = 1, \ldots, m$). There has to be considered the admissibility on these data of the hypothesis that the agents transformations are subject to some transformation-possibility relation T which is on the normal model, that is which satisfies the six normality axioms. Then further, in case of admissibility, which case can define normal consistency for the transformation data, it is required to characterize all possible normal relations which could be admitted as hypothesis.

The rth observed transformation can be stated as the transformation displacement $z_t = y_t - x_t$ applied to the stock of goods x. The fundamental theorem on normality easily shows that normal consistency is equivalent to the condition that there exists $u > 0$ such that $u'z_t \leq 0$ ($t = 1, \ldots, m$). With this condition satisfied, $U = \{u \geq 0; u'z_t \leq 0$ ($t = 1, \ldots, m$) is a polyhedral convex cone cutting the interior of Ω. Let $u_1, \ldots, u_k$ be a fundamental set of solutions of $u'z_t \leq 0$, $u \geq 0$ ($t = 1, \ldots, m$), and let T be the relation defined by $xTy \Leftrightarrow u_r'x \leq u_r'y$ ($r = 1, \ldots, k$), $x, y \geq 0$. Then T is a normal transformation-possibility relation satisfying the admissibility requirement $x_t T y_t$ ($t = 1, \ldots, m$), and a necessary and sufficient condition that any other normal relation T* satisfy this requirement is $T \subseteq T^*$. Thus T appears as the minimal admissible normal relation. For associated cones, the corresponding relation is $U \supseteq U^*$. The normal consistency condition for data is now established, and all admissible normal relations have been characterized in case it is

satisfied. This extremely simple principle can be elaborated to give
different general methods for the empirical construction of input-output
relations, in particular of production functions, the methods differing
according to the form, especially the particular mode of incompleteness,
of the data. In particular, consumers' utility functions can be con-
structed from budget data,[15] which is a peculiarly incomplete form of data
since the quantity of output is not given, and instead, prices are given.
Then it is possible to proceed further, and give construction of cost-
of-living indices, determined from the utility functions. This method
for the consumer has already been sytematically and extensively explored.[16]
It leads to a general view of consumer index-number questions, and a
theory of index-number construction from budget data which is entirely
satisfactory within that view, and has evident availability for practical use. The method of Wald,[17] together with a substantial generalization
of that method,[18] can be placed directly into the framework of the general
approach, where it can be given a necessary qualification, and then a
simplification, and, through relation to the general approach, a fuller
interpretation. The practical need for developed methods of index-
number construction is a clear recommendation of the Stigler Committee.[19]
Most practical effort at present goes towards the question of fixing
weights in the conventional index. But weighting is itself an inadequate concept; and it is not clear what is gained in the refinement
of an inadequate concept. Returning again to production theory, development of these empirical methods[20] gives finite computation together
with the flexibility required for incorporating important structural

features which cannot be represented by simple production functions,
or in case they can, then not by any of those few models which have
actually been put to empirical use, such as the homogeneous classical
type Cobb-Douglas production function, or the Leontief input-output
system.

There is an analogy between the axiomatic analysis for normal
transformation-possibility which has been given here, and the well-
known axiomatic analysis of von Neumann-Morgenstern utility.[21] Moreover, an empirical approach can be made to utility which is analogous
to that for transformation-possibility. An account of that approach,
which states the most general approach to measurement of von Neumann-
Morgenstern utility, has been given elsewhere.[22] In that account, behavior data is assumed given which will reveal relations of preference
between certain probability distributions over possible events. The
data, or more specifically, the preference relations, are assumed finite,
so as to permit a finite algebraical treatment, but in principle there
need be no such restriction. The consistency condition on the data, in
order that the utility hypothesis be admissible, is shown. Then, in
case it is satisfied, all those utility functions which are possible
hypotheses are characterized, as belonging to a certain convex cone.
With the finiteness assumption, it is a polyhedral cone. In case the
data set is enlarged, while remaining consistent, the cone narrows; and
if it is enlarged to completeness, the cone narrows to a ray, that is,
the admissible utility function becomes essentially unique. But
generally, the cone of utility functions determines a relation between

probability distributions which satisfies all the utility axioms but
the axiom of completeness. Also Aumann[23] has considered utility in the
absence of the completeness assumption.

The quantitative investigation of economic relationships, by means
of algebraical models in the form considered here, has an appropriate-
ness which is peculiar and intrinsic to the nature of economic quantities,
in contrast to the general statistical methods, which have no such special
affinity, but which have been the main source of econometric methods.
Nevertheless these means have been scarcely considered. The Leontief
input-output system is a special instance. Another is the previously
mentioned construction of consumers' utility functions from finite, consistent expenditure data, and also the measurement of von Neumann-
Morgenstern utility. There seems to be no other examples which are
familiar. Recently Scarf[24] has shown how equilibrium prices for an exchange economy can be finitely calculated, if utility functions are
assumed, as in the form of the above-mentioned construction, equal to
the minimum of a finite set of linear functions. This is a significant
step towards freeing an essentially finite linear-algebraical subject
from any of the usual reliance on topological methods. Linear formulation is not a limitation which precludes the convexity-concavity
assumptions which are characteristic of much economic theory. Convexity
is a linear concept; and, by approximation, it is a finite linear concept. In fact, all the general theorems of concave programming can be
deduced from the standard theorems of linear programming. The deduction
is direct, in case of polyhedral objective and constraint functions;

otherwise, a limiting process is required. The convexity assumptions which are an important part of the foundation of economic theory can themselves be taken as having a foundation in the normal axioms of transformation-possibility.

Notes

(q(p): note q, appearing on page p)

1(2):   Koopmans [14].

2(2):   von Neumann [17].

3(3):   Concerning the static character of Koopmans' model, and
          static and dynamic models in general, see Koopmans [17], p.
          35, cf. Georgescu-Roegen [11], pp. 100-101.

4(6):   This is the basic assumption of the von Neumann model.

4a(6):   Karlin [13], p. 403.

5(7):   The one is explicit about possible transformations, and the
          other, about impossible.

5a(7):   Gale [10], p. 56.

6(8):   Koopmans' postulate on "the impossibility of the Land of
          Cockaigne", Koopmans [14], p. 49.

6a(8):   Fenchel [8], p. 48.

6b(8):   Rather, the argument shows V has a dual U. Now Ω, $-\Omega$ are
          dual. Hence, from $V \supset -\Omega$, by taking duals, it follows that
          $U \subseteq \Omega$. Then further, U cannot be contained in the boundary
          of Ω, for this would lead to a contradiction of $V\mathfrak{m} = \{o\}$.
          Accordingly, U lies in, and cuts the interior of Ω. (Strict-
          ly, U lies in a replica of Ω which, for simplicity, can be
          identified with Ω.)

6c(9):   With $U \subseteq \Omega$ must be joined the further remark that U cuts
          the interior of Ω. This applies also in the next paragraph,
          and note 13a(26).

7(10):   cf. Drandakis [7], p. 333.

8(10):   Usages for relations, here and elsewhere in this paper,
          follow Afriat [1]. Thus, with $A,B \subseteq S{\times}S$, $AB = (x,y):(Vz)$
          $(x,z) \in A$, $(z,y) \in B$.

9(10):   von Neumann [17].

10(11):   cf. Drandakis [7], p. 333.

11(23):   cf. Debreu [6], p. 42.

11a(23): $x\bar{T}y$ denotes not xTy.

12(23):   cf. Postulate B, Koopmans [14], p. 50.

13(24):   cf. Georgescu-Roegen [11], p. 103.

13a(26): See note 6c(9).

14(38):   cf. Georgescu-Reogen [11], p. 99.

15(48):   Afriat [3].

16(48):   Afriat [4] outlines a general approach to index construction, in which solution of a certain system of inequalities $\lambda_r \geq$ o, $\lambda_r D_{rs} \geq \Phi_s - \Phi_r (r,s = 1,\ldots,k)$, has a fundamental place. The coefficients $D_{rs} = p_r' x_s / p_r' x_r - 1$ are formed from price-quantity data (p_r, x_r) observed on k occasions. Even such special methods as involve quadratic utility functions can be placed, and interpreted, within this general approach. The theory can be assembled more fully out of Research Memoranda Nos. 21, 24 and 27 (1961), Econometric Research Program, Princeton University (unpublished). Certain problems which arise are treated in "The system of inequalities $a_{rs} > X_s - X_r$," Proc. Cambridge Phil. Soc., 59(1963), 125-133, and "Gradient configurations and quadratic functions," Proc. Cambridge Phil. Soc., 59(1963), 287-305. The classical case of two-period data (k=2) is treated in "The method of limits in the theory of index-numbers" (unpublished). An observation concerning this case is contained in "An identity concerning the relation between the Paasche and Laspeyres indices," Metroeconomica XV(1963), II-III, 136-140.

17(48):   Wald [19], Afriat [4].

18(48):   Afriat [4] shows the nature of this generalization, and the algorithm for it, together with a computer program and numerical and graphical illustrations. Formal derivation is in Research memorandum No. 24 (see note 16), and "Gradient configurations ..." (note 16).

19(48):   Dr. George J. Stigler, Chairman, Price Statistics Review, Committee of the National Bureau of Economic Research. U. S. Congress Joint Economic Committee [16] presents the report.

20(48):   That is, based on the normal model for a transformation-possibility relation, that is, virtually the Koopmans activity model.

21(49):   von Neumann-Morgenstern [18].

22(49):   Afriat [2].

23(50):   Aumann [5].

24(50):   Scarf [15].

Figures

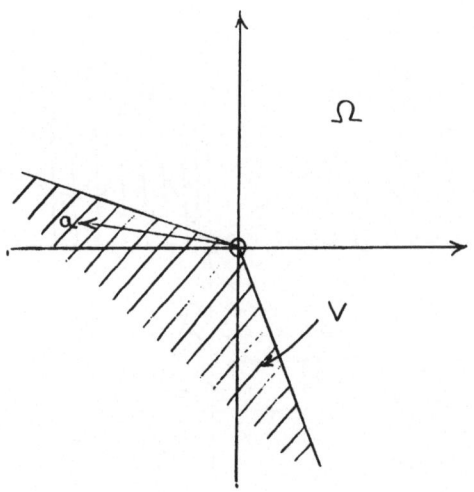

Fig. 1 (p.5)

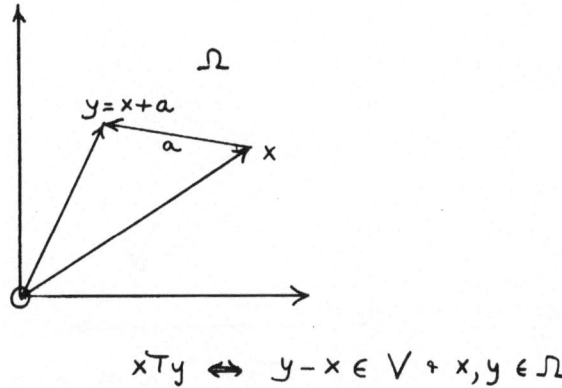

$$xTy \iff y-x \in V \cdot x, y \in \Omega$$

Fig. 2 (p.5)

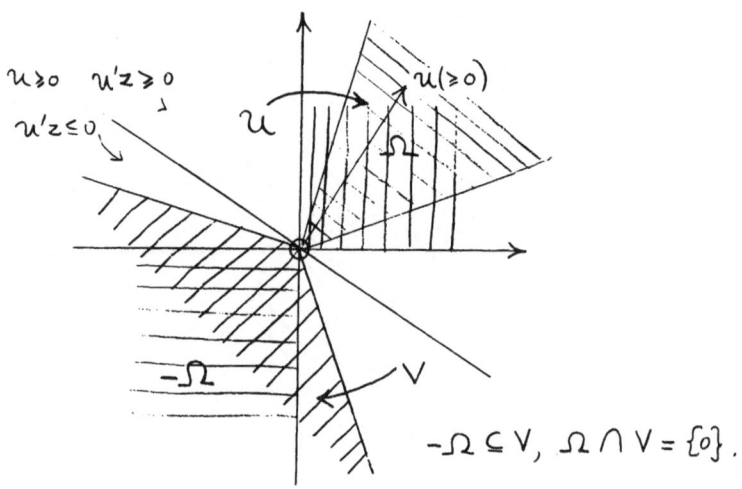

Fig. 3 (p. 10)

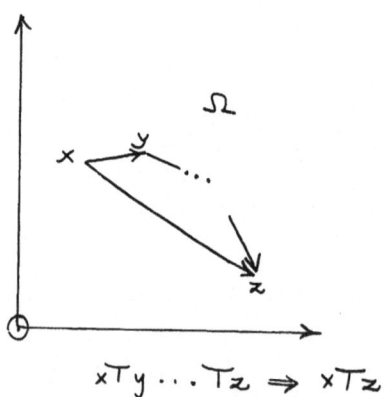

Fig. 4 (p. 18)

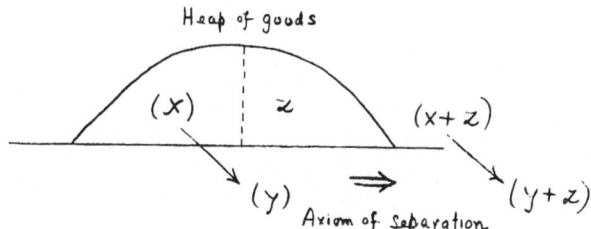

Fig. 5a (p. 18)

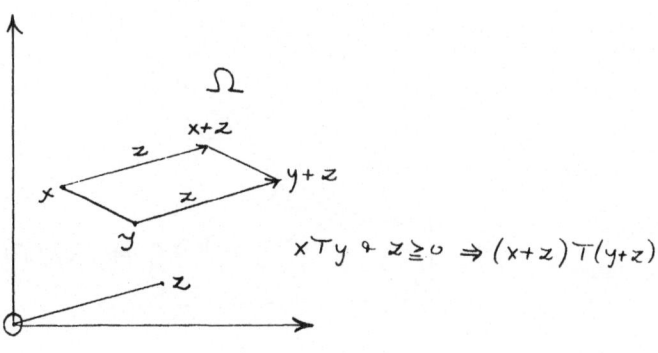

Fig. 5b (p. 18)

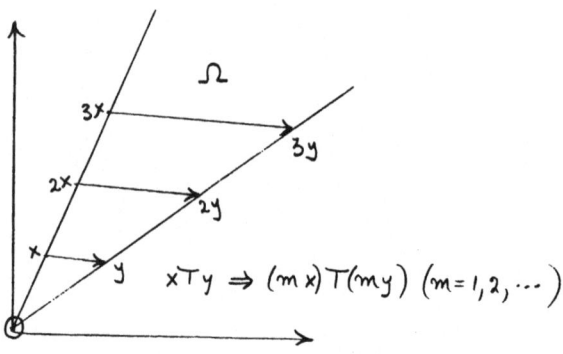

Fig. 6 (p. 20)

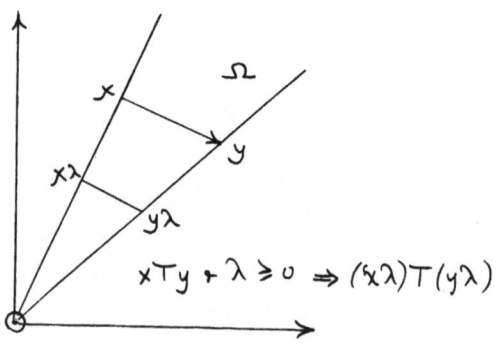

Fig. 7 (p. 20)

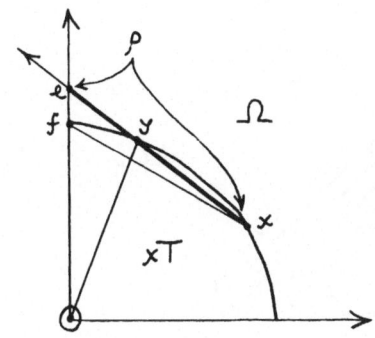

Fig. 8a (p. 33)

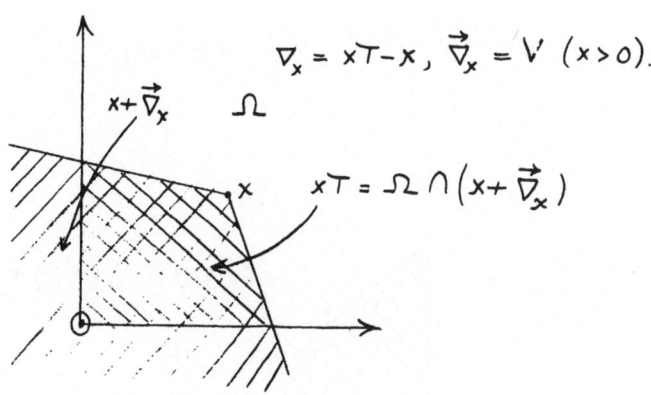

Fig. 8b (p. 33)

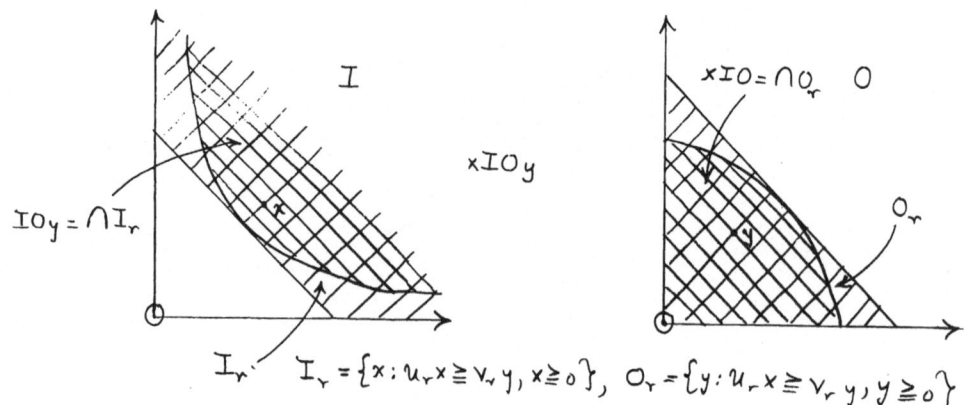

$$I_r = \{x : \mathcal{U}_r x \geqq V_r y, x \geqq o\}, \quad O_r = \{y : \mathcal{U}_r x \geqq V_r y, y \geqq o\}$$

Fig. 9 (p. 39)

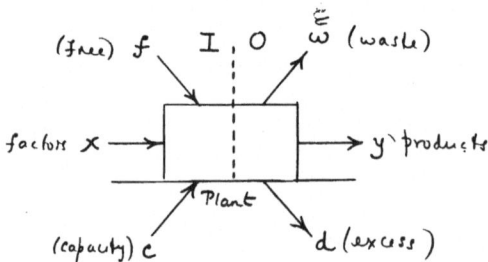

Fig. 10 (p. 44)

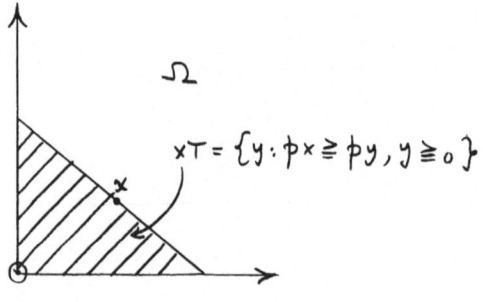

$$xT = \{y : px \geqq py, y \geqq o\}$$

Fig. 11 (p. 46)

References

[1]   Afriat, S. N.: "Preference Scales and Expenditure Systems,"
      Econometrica 30(1962), 305-323.

[2]   Afriat, S. N.: "The Validity of the Expected Utility Hypothesis,"
      Recent Advances in Game Theory, Princeton, 1962, 73-82.

[3]   Afriat, S. N.: "The Construction of Utility Functions from Ex-
      penditure Data," Cowles Foundation Discussion Paper No. 144
      (October, 1964), to appear in the International Economic Review.

[4]   Afriat, S. N.: "The Cost of Living Index," M. Shubik (Ed.), Studies
      in Mathematical Economics: Essays in Honor of Oskar Morgenstern,
      (Chapter 13), to be published by the Princeton University Press.

[5]   Aumann, R. J.: "Utility Theory without the Completeness Axiom,"
      Econometrica 30, 3(1962), 445-462.

[6]   Debreu, G.: Theory of Value, Cowles Foundation Monograph No. 17,
      Wiley, New York, 1959.

[7]   Drandakis, E. M.: "On Efficient Accumulation Paths in the Closed
      Production Model," Econometrica 34, 2(1966), pp. 331-346.

[8]   Fenchel, W.: Convex Cones, Sets and Functions, (Notes of lectures
      at Princeton University, Spring Term, 1951) Department of Math-
      ematics, Princeton University, 1953.

[9]   Gale, D.: "The Closed Linear Model of Production," H. W. Kuhn and
      A. W. Tucker (Eds.), Linear Inequalities and Related Systems,
      Annals of Mathematics Studies No. 38, Princeton, 1956.

[10] Gale, D.: The Theory of Linear Economic Models, McGraw-Hill,
      New York, 1960.

[11] Georgescu-Roegen, N.: "The Aggregate Linear Production Function
      and its Applications to the von Neumann Model," T. C. Koopmans
      (Ed.), Activity Analysis of Production Allocation (Chapter IV),
      Cowles Commission Monograph No. 13, New York, 1951.

[12] Goldman, A. J. and A. W. Tucker: "Polyhedral Convex Cones," H. W.
      Kuhn and A. W. Tucker, (Eds.), Linear Inequalities and Related
      Systems, Annals of Mathematics Studies No. 38, Princeton, 1956.

[13] Karlin, S.: Mathematical Methods and Theory in Games, Programming
      and Economics, Addison-Wesley, 1959.

[14] Koopmans, T. C.: "Analysis of Production as an Efficient Combination
      of Activities," Activity Analysis of Production and Allocation,
      Ed. T. C. Koopmans, Wiley, New York, 1951, Chapter III, pp. 33-97.

[15] Scarf, H.: "Equilibrium Points of Two Person Games and the
      Brouwer Fixed Point Theorem," unpublished manuscript.

[16] U. S. Congress, Joint Economic Committee: Government Price
      Statistics, Hearings, January 24, 1961, GPO, Washington 25, D.C.

[17] von Neumann, J.: "A Model of General Economic Equilibrium,"
      Review of Economic Studies, 13(1945-46), pp. 1-9.

[18] von Neumann, J. and O. Morgenstern: Theory of Games and Economic
      Behaviour, 3rd Ed., Princeton, 1953.

[19] Wald, A.: "A New Formula for the Index of Cost of Living,"
      Econometrica, 7, 4(1939), pp. 319-335.

CENTRO INTERNAZIONALE MATEMATICO ESTIVO

(C. I. M. E.)

M. ARCELLI

"MODELLI AUMENTATI E PRINCIPIO DI CORRISPONDENZA
NELLA METODOLOGIA DI ANDREAS G. PAPANDREOU"

Corso tenuto a Villa Falconieri (Frascati) dal 22 al 30 agosto 1966

MODELLI AUMENTATI E PRINCIPIO DI CORRISPONDENZA
NELLA METODOLOGIA DI ANDREAS G. PAPANDREOU

di

M. Arcelli

Con la presente relazione mi propongo di illustrare taluni svi-
luppi della metodologia economica di Andreas G. Papandreou [1].
Tema centrale dell'analisi metodologica del Papandreou è la statica comparata , il cui significato, come è stato posto in evidenza dal Samuelson, si completa attraverso un'intima relazione formale con la dinamica. Se non si può negare a Samuelson [2] il primato di aver rigoro-
samente definito le connessioni tra statica comparata e dinamica attraverso il principio di corrispondenza, non si può disconoscere al Papandreou
il merito di aver chiarito il contenuto epistemologico di tale rapporto.
Lo svolgimento di questo argomento forma l'oggetto della presente esposizione . Il discorso, tuttavia, deve iniziare più a monte, partendo dai
problemi più generali di interpretazione empirica dei termini e delle
proposizioni economiche, per risalire gradualmente alla comprensione
del significato operativo dei teoremi di statica comparata .

Abbandonata l'ingenua illusione di poter definire completamente
in termini d'osservabili ogni termine delle proposizioni scientifiche ;
respinto il credo secondo cui ogni enunciato potrebbe venir tradotto, mediante opportune definizioni dei suoi termini costitutivi in un enunciato
equivalente, espresso in termini empirici; rimangono acquisiti alla mo-
derna metodologia scientifica il rigore e la coerenza del linguaggio che
inducono a rafforzare l'assiomatizzazione dei sistemi di proposizioni, e

---

[1] Mi riferisco in modo preminente al volume, L'Economia come scienza,
L'industria, Milano, 1962, da me tradotto ed annotato.

[2] Si veda Samuelson P . , Foundations of Economic Analysis, Harvard
University Press, 1961 (6^a edizione)

M. Arcelli

a ricercare nel collegamento globale delle ipotesi con la realtà, feconde connessioni ai fini della spiegazione e della previsione. Nella scienza economica, il collegamento delle ipotesi con la realtà, non potrà naturalmente ignorare le complicazioni derivanti dal condizionamento storico [3]

L'impossibilità di definire direttamente ed indipendentemente ogni termine delle proposizioni economiche, è imputabile all'esistenza di termini disposizionali e, nelle proposizioni a più elevato grado di universalità, di termini teorici.

I termini disposizionali non definiscono proprietà autonome, direttamente osservabili, bensì comportamenti che implicano particolari reazioni in determinate circostanze. Così un investimento è inflazionistico se la domanda globale supera l'offerta globale ; è propulsivo nel caso opposto, quando esistono fattori disoccupati. Inflazionistico e propulsivo sono termini disposizionali, definibili solo parzialmente, poichè gli stessi termini sono pure definibili in altri contesti.

I termini disposizionali sono quindi traducibili in concetti operativi, limitatamente alle situazioni che soddisfano particolari "condizioni osservabili" (3/bis) .

I termini teorici (ad es . : moltiplicatore, acceleratore ecc.) non sono definibili direttamente, neppure condizionalmente. Essi ricevono una definizione operativa indirettamente, attraverso l'interpretazione empirica globale della teoria in cui sono incorporati e da cui non possono essere eliminati senza che la stessa risulti mutilata.

---

[3] Si veda nella nostra introduzione a l'economia come scienza, op. cit. , p. 23 il paragrafo dedicato alla definizione del concetto di spazio sociale e le conseguenze di una mancata identificazione dello stesso sullo "status scientifico" delle uniformità economiche .

(3/bis) ibidem p. 21

M.Arcelli

E'percio indispensabile che le proposizioni economiche, ordina-
te in sistemi, ricevano un'interpretazione empirica globale, una volta
verificata la loro coerenza logica, attraverso un rigoroso processo di assiomatizzazione.

Qualunque processo di assiomatizzazione si traduce, almeno in
un primo tempo, nella separazione di ciò che vi è di formale, da ciò
che costituisce il contenuto specifico del sistema di proposizioni sotto-
posto ad analisi. Il contenuto si rivela estrinseco rispetto alla forma.
Quest'ultima si manifesta nei segni e nella struttura, ossia nelle relazioni che collegano un termine ad un altro, una proposizione ad un'altra.

Il contenuto, tuttavia, come constateremo a proposito delle
variabili incluse nelle proposizioni economiche, condiziona la forma della
struttura, imponendo vincoli o assegnando una direzione al verso delle
relazioni.

In economia, il concetto di struttura quale emerge dall'analisi logica si identifica quindi con quello di insieme di relazioni consistenti
che collegano una serie di termini o di proposizioni appartenenti ai sistemi economici. Le relazioni in sè, non possiedono significato. La loro
interpretazione empirica è il fondamento della successiva spiegazione
scientifica.

Limitando la considerazione a quella parte dell'economia esprimi-
bile, in termini matematici, la struttura è definibile come un'inte-
laiatura referenziale denotata da un sistema d'equazioni consistenti ed
indipendenti. Se si adottasse una simile definizione, senza alcun vincolo
empirico, la discussione di tale concetto potrebbe ricondursi a quella
dei sistemi matematici. Volendo differenziare le strutture e le variabili economiche dagli altri tipi di struttura e di variabile è necessario
superare le proprietà formali generali, con la considerazione delle ca-

M. Arcelli

catteristiche distintive. Ciò si ottiene mediante l'interprezazione empirica [4] . Si tratta di valutare il rapporto con la realtà economica dei
termini della struttura. Ove non sia possibile una definizione diretta e
completa di tutti i termini, con riferimento alla realtà empirica, l'interpretazione risulterà indirettamente : quest'ultimo caso vale per i termin
teorici e per le ipotesi a più alto livello di universalità. Reddito, invest
menti, consumi, prezzi , costi ecc. sono termini economici rappresenta
dalle X, Y, Z,   ecc. dei sistemi di equazioni. Essendo l'economia una
scienza empirica, i correlati alle X, Y, Z devono essere tratti dal
tessuto empirico dell'esperienza economica, cioè devono risultare da
particolari tipi d'osservazione. Il campo di variazione delle variabili eco
nomiche e quindi le caratteristiche delle strutture economiche trovano
pertanto dei limiti ben definiti nell'osservazione delle realtà economica.
L'interpretazione empirica permette di chiudere gli enunciati aperti
rappresentati dalle equazioni che soddisfano le relazioni da esse definite.
Prendiamo un esempio concreto. Sia il sistema [5] .

1) $Y = C + I + G$

2) $Z = Y - T$

3) $C = a + bZ$

4) $I = u + vY$

5) $T = r Y$

---

[4] Si veda la nostra introduzione a, l'economia come scienza, op. cit. ,
    p. 19.

[5] Esso è tratto da un'appendice al volume del Papandreou, l'Economia
come scienza, op. cit., La sua successiva dinamizzazione sarà invece frut-
to di nostre elaborazioni.

M. Arcelli

che scritto in forma implicita diviene

1) $Y - C - I - G = 0$

2) $Y - Z - T = 0$

3) $C - a - bZ = 0$

4) $I - u - vY = 0$

5) $T - rY = 0$

ed in forma generica corrispondente alla forma implicita

$$\phi = \begin{cases} F^1(Y, \ Z, \ C, \ I, \ T, \ G) = 0 \\ F^2(Y, \ Z, \ C, \ I, \ T, \ G) = 0 \\ F^3(Y, \ Z, \ C, \ I, \ T, \ G) = 0 \\ F^4(Y, \ Z, \ C, \ I, \ T, \ G) = 0 \\ F^5(Y, \ Z, \ C, \ I, \ T, \ G) = 0 \end{cases}$$

Tale sistema è un sistema matematico costituito da equazioni , cioè
da enunciati aperti. Esso assume le caratteristiche di struttura economica solo in seguito ad interpretazione empirica delle variabili, sul fondamento di atti d'osservazione della realtà economica. In tal modo si
realizza la chiusura degli enunciati, che si trasformano in proposizioni.

Se indichiamo con $0 = \{0_1, 0_2, \ldots, 0_n\}$ una classe d'atti d'osservazione che individuano X (dove con X indichiamo sinteticamente tutte
le variabili del nostro sistema), ad ogni atto d'osservazione 0 corri-
spondono determinati valori delle variabili denotate sinteticamente da X.
Le nostre equazioni, esprimenti ora proposizioni economiche, saranno
definite esclusivamente da quegli atti d'osservazione che soddisfano le
relazioni F^i , cioè , in termini di logica simbolica , da

$$\bigwedge_0 \ \bigwedge_X \ \left\{ (0, \ X) \varepsilon \ \ 0 \rightarrow X \, \varepsilon \, F^i \right\}$$

Tale espressione si legge : per tutti gli atti d'osservazione 0 e per
tutti gli X, se X è osservato secondo la regola 0; allora soddisfa

la relazione F^i .

In tal modo abbiamo tradotto il sistema matematico in struttura economica o , con le parole del Papandreou, abbiamo fornito un'interpretazione di base alle proposizioni del nostro sistema. Stabiliamo pertanto una correlazione tra atti d'osservazione e valori delle variabili,
convenendo d'indicare con   Y   il reddito nazionale, con Z il reddito disponibile, con C   i consumi, con I   gli investimenti, con   T
il   gettito fiscale di un'imposta sul reddito e con G   le spese governative. Le lettere minuscole definiscono parametri. Il sistema è   di tipo
keynesiano, con l'introduzione esplicita di un meccanismo fiscale e con
l'eliminazione dell'equazione monetaria. Delle sei variabili viene scelta
come esogena la variabile spese governative, secondo una tradizione or-
mai consolidata.

In presenza di un tale sistema, l'economista si propone generalmente i seguenti scopi, che si completano reciprocamente :

     a) Verificare il   modello. (Il sistema determina i valori d'equili-
        brio che bisogna interpretare empiricamente, riscontrandone o
        o meno l'esistenza nella concreta realtà economica) ;

     b) Derivare teoremi di statica comparata. L'economista non si
        accontenta di determinare l'equilibrio . Vuole sapere cosa
        succede scostandosi dall'equilibrio) .

Entrambi gli scopi (a e b) richiedono la formazione di sistemi
dinamici corrispondenti al sistema statico da analizzare. I valori d'equilibrio infatti, devono interdersi come valori stazionari. Se accettiamo la
definizione di Samuelson [6] per cui un sistema dinamico è un insieme di
equazioni funzionali che, con le condizioni iniziali (nel senso più genera-

---

[6] Si veda Saumelson P. , Foundations of Economic Analysis, op. cit.,
p. 284 .

M. Arcelli

rale), determina come soluzioni certe incognite in funzione del tempo ;
allora i sistemi statici sarebbero semplicemente casi speciali, degene-
rati, in cui le equazioni funzionali assumono una forma semplificata e
determinano come soluzioni funzioni di tempo che sono identicamente
costanti : onde il concetto di valori d'equilibrio o soluzioni stazionarie.
Oppure possono essere intesi come limite di un sistema dinamico smor-
zato. D'altro lato, se consideriamo l'economia qualitativa, che è quella
che si rivela maggiormente verificabile empiricamente, in essa i teoremi di statica comparata definiscono il segno della variazione di variabili
endogene in risposta a mutamenti intervenuti in una o più variabili
esogene. Ma tale segno non sempre può essere derivato in modo certo,
esclusivamente sul fondamento di un sistema statico. Generalmente è es-
senziale che l'equilibrio del sistema sia stabile : il concetto di stabilità ,
tuttavia, può essere inteso solo con riferimento ad un sistema dinamico
che incorpori le ipotesi del sistema statico di base. Le condizioni di
stabilità ricavate da tale analisi, sono spesso strumenti indispensabili
per la risoluzione di teoremi di statica comparata : la stabilità qui
intesa, coincide con la stabilità in senso matematico delle equazioni
costituenti i sistemi dinamici. Conversamente le proprietà del sistema
statico concedono informazioni relative al comportamento di un sistema
dinamico corrispondente. Questa relazione bidirezionale tra statica e dinamica costituisce il principio di corrispondenza di Samuelson . Come affer-
ma Morishima [7] , riferendosi però in particolare ai problemi dell'equilibrio di Walras-Leontief, il principio di corrispondenza di Samuelson tra
statica e dinamica mostra come il problema di derivare teoremi di statica comparata, significativi operativamente, sia strettamente legato al pro-

---

[7]Morishima M., Equilibrium stability and growth, Oxford University Press,
1964, p. 24)

blema della stabilità dell'equilibrio .

Poichè , tuttavia, tale principio enuncia la relazione tra stabilità dell'equilibrio nel piccolo e cambiamenti nella posizione d'equilibrio dovuti, ad un piccolo mutamento di un dato, potrebbe denominarsi principio di corrispondenza nel piccolo . In altri termini , non si prendono in considerazione nel nostro studio i problemi della stabilità globale, anche se può mostrarsi che diversi teoremi di statica comparata di carattere non infinitesimale mostrano un'intima connessione con le proprietà della stabilità globale del sistema.

In sostanza, come sostiene il Papandreou, il processo di verifica di un modello statico o di una proposizione di economia qualitativa che definisce un teorema di statica comparata, richiede un ampliamento ed una revisione del modello in modo da permettere tale verifica. Si deve passare cioè da un'interpretazione di base ad un'interpretazione aumentata, da un modello di base statico ad un modello aumentato dinamico. Non esiste tuttavia generalmente un'unica interpretazione aumentata corrispondente ad una data interpretazione di base : l'economista può cioè costruire diversi modelli dinamici alternativi corrispondenti a un dato sistema di base. In questo senso, conclude il Papandreou [8] "la teoria sarà respinta solo se tutti gli elementi di una classe di interpretazioni aumentate portano all'invalidazione delle ipotesi della teoria, come risultato del confronto con i dati empirici".

Prendendo spunto dal sistema composto dalle cinque equazioni, cui abbiamo fornito un'interpretazione di base, ci proponiamo ora di illustrare concretamente i problemi metodologici esposti, attraverso la verifica di proposizioni qualitative di statica comparata. Pur riferendo-

---

[8] Si veda L'economia come scienza, op. cit. , p. 141

M. Arcelli

ci ad un modello particolare cercheremo tuttavia di generalizzare alquan-
to i risultati, considerando anche il modello in forma generica.

Supponiamo dunque di volere determinare il verso di variazione
delle variabili endogene, in corrispondenza di un mutamento della variabile esogena G.
Per ottenere ciò potremo esprimere le equazioni del sistema in forma ridotta e quindi derivare la variabile endogena di ogni equazione rispetto
alla variabile G. Otterremmo così :

$$\left(\frac{\delta Y}{\delta G}\right)^{\circ} = \frac{1}{1-[b(1-\imath)+\upsilon]} \; ; \; \left(\frac{\delta Z}{\delta G}\right)^{\circ} = \frac{1-\imath}{1-[b(1-\imath)+\upsilon]} \; ; \; \left(\frac{\delta C}{\delta G}\right)^{\circ} = \frac{b(1-\imath)}{1-[b(1-\imath)+\upsilon]} \; ;$$

$$\left(\frac{\delta I}{\delta G}\right)^{\circ} = \frac{\upsilon}{1-[b(1-\imath)+\upsilon]} \; ; \left(\frac{\delta T}{\delta G}\right)^{\circ} = \frac{\imath}{1-[b(1-\imath)+\upsilon]}$$

Infatti , ad es. ,

$$Y = \frac{\alpha + u}{1-[b(1-\imath)+\upsilon]} + \frac{G}{1-[b(1-\imath)+\upsilon]} \qquad \text{da cui} \qquad \left(\frac{\delta Y}{\delta G}\right)^{\circ} = \frac{1}{1-[b(1-\imath)+\upsilon]}$$

Tuttavia non sempre è agevole passare alla forma ridotta, non solo :
tale trattamento manca di generalità rispetto al seguente procedimento che
ci permetterà di seguire lo sviluppo dei passaggi matematici contemporanea-
mente sul sistema in forma implicita ed in forma generica .

Sia $\overset{\circ}{X} = \left\{\overset{\circ}{Y}, \; \overset{\circ}{Z}, \; \overset{\circ}{C}, \; \overset{\circ}{I}, \; \overset{\circ}{T}, \; \overset{\circ}{G}\right\}$ una soluzione di ϕ

Se $F^{1}, F^{2} \dots F^{5}$ sono continue e derivabili e se in $\overset{\circ}{X}$ lo Jacobiano
è non singolare

M. Arcelli

$$J = \begin{vmatrix} F^1_y & F^1_z & F^1_c & F^1_I & F^1_T \\ F^2_y & F^2_z & F^2_c & F^2_I & F^2_T \\ F^3_y & F^3_z & F^3_c & F^3_I & F^3_T \\ F^4_y & F^4_z & F^4_c & F^4_I & F^4_T \\ F^5_y & F^5_z & F^5_c & F^5_I & F^5_T \end{vmatrix} \neq 0 \text{ cioè } \Delta = \begin{vmatrix} Y & Z & C & I & T \\ 1 & 0 & -1 & -1 & 0 \\ 1 & -1 & 0 & 0 & -1 \\ 0 & -b & 1 & 0 & 0 \\ -V & 0 & 0 & 1 & 0 \\ -r & 0 & 0 & 0 & 1 \end{vmatrix} \neq 0$$

allora esiste un intorno di $\overset{\circ}{G}$ in cui le variabili endogene sono funzioni univoche e differenziabili di G, cioè è possibile scrivere, derivando totalmente rispetto a G : [9]

$$\begin{cases} F^1_Y\left(\frac{\delta Y}{\delta G}\right)^0 + F^1_z\left(\frac{\delta Z}{\delta G}\right)^0 + F^1_c\left(\frac{\delta C}{\delta G}\right)^0 + F^1_I\left(\frac{\delta I}{\delta G}\right)^0 + F^1\left(\frac{\delta T}{\delta G}\right)^0 = -F^1_G \\ F^2_Y\left(\frac{\delta Y}{\delta G}\right)^0 + F^2_z\left(\frac{\delta Z}{\delta G}\right)^0 + F^2_c\left(\frac{\delta C}{\delta G}\right)^0 + F^2_I\left(\frac{\delta I}{\delta G}\right)^0 + F^2\left(\frac{\delta T}{\delta G}\right)^0 = -F^2_G \\ F^3_Y\left(\frac{\delta Y}{\delta G}\right)^0 + F^3_z\left(\frac{\delta Z}{\delta G}\right)^0 + F^3_c\left(\frac{\delta C}{\delta G}\right)^0 + F^3_I\left(\frac{\delta I}{\delta G}\right)^0 + F^3\left(\frac{\delta T}{\delta G}\right)^0 = -F^3_G \\ F^4_Y\left(\frac{\delta Y}{\delta G}\right)^0 + F^4_z\left(\frac{\delta Z}{\delta G}\right)^0 + F^4_c\left(\frac{\delta C}{\delta G}\right)^0 + F^4_I\left(\frac{\delta I}{\delta G}\right)^0 + F^4\left(\frac{\delta T}{\delta G}\right)^0 = -F^4_G \\ F^5_Y\left(\frac{\delta Y}{\delta G}\right)^0 + F^5_z\left(\frac{\delta Z}{\delta G}\right)^0 + F^5_c\left(\frac{\delta C}{\delta G}\right)^0 + F^5_I\left(\frac{\delta I}{\delta G}\right)^0 + F^5\left(\frac{\delta T}{\delta G}\right)^0 = -F^5_G \end{cases}$$

dove gli F^i_{xk} sono pienamente determinati nel punto di equilibrio $\overset{\circ}{X}$ e rappresentano i coefficienti, mentre $\left(\frac{\delta x_k}{\delta G}\right)^0$ sono le incognite di un sistema lineare non omogeneo di 5 equazioni in 5 Variabili. Nel nostro caso avremo pertanto :

$$\left(\frac{\delta Y}{\delta G}\right)^0 - \left(\frac{\delta C}{\delta G}\right)^0 - \left(\frac{\delta I}{\delta G}\right)^0 = \cdot 1$$

---

[9]

Ciò risulta da teoremi sulle funzioni implicite.

M.Arcelli

$$
\begin{cases}
\left(\dfrac{\delta Y}{\delta G}\right)^{\!o} - \left(\dfrac{\delta Z}{\delta G}\right)^{\!o} - \left(\dfrac{\delta T}{\delta G}\right)^{\!o} = 0 \\[2em]
-b\left(\dfrac{\delta Z}{\delta G}\right)^{\!o} + \left(\dfrac{\delta C}{\delta G}\right)^{\!o} = 0 \\[2em]
-v\left(\dfrac{\delta Y}{\delta G}\right)^{\!o} + \left(\dfrac{\delta I}{\delta G}\right)^{\!o} = 0 \\[2em]
-\eta\left(\dfrac{\delta Y}{\delta G}\right)^{\!o} + \left(\dfrac{\delta T}{\delta G}\right)^{\!o} = 0
\end{cases}
$$

Applicando la regola di Cramer, la generica formula risolutiva risulta

$$
\left(\frac{\delta X_k}{\delta G}\right)^{\!o} = - \frac{\sum F_G^i \, \Delta_{ik}}{\Delta}
$$

Ne segue che :

$$
\left(\frac{\delta Y}{\delta G}\right)^{\!o} = \frac{\Delta_{11}}{\Delta} \; ; \; \left(\frac{\delta Z}{\delta G}\right)^{\!o} = \frac{\Delta_{12}}{\Delta} \; ; \; \left(\frac{\delta C}{\delta G}\right)^{\!o} = \frac{\Delta_{13}}{\Delta} \; ; \; \left(\frac{\delta I}{\delta G}\right)^{\!o} = \frac{\Delta_{14}}{\Delta} \; ; \; \left(\frac{\delta T}{\delta G}\right)^{\!o} = \frac{\Delta_{15}}{\Delta}
$$

e poichè $\Delta = -1 + [b(1-r) + V]$; $\Delta_{11} = -1$; $\Delta_{12} = -(1-r)$;

$\Delta_{13} = -b(1-r)$; $\Delta_{14} = -V$; $\Delta_{15} = -r$

$$
\left(\frac{\delta Y}{\delta G}\right)^{\!o} = \frac{1}{1-[b(1-\eta)+v]} \; ; \; \left(\frac{\delta Z}{\delta G}\right)^{\!o} = \frac{1-\eta}{1-[b(1-\eta)+v]} \; ; \; \left(\frac{\delta C}{\delta G}\right)^{\!o} = \frac{b(1-\eta)}{1-[b(1-\eta)+v]} \; ; \; \left(\frac{\delta I}{\delta G}\right)^{\!o} = \frac{v}{1-[b(1-\eta)+v]} \; ;
$$

$$
\left(\frac{\delta T}{\delta G}\right)^{\!o} = \frac{\eta}{1-[b(1-\eta)+v]}
$$
; cioè come era logico attendersi, gli stessi
risultati ottenuti esprimendo il sistema in forma ridotta e derivando. Se
avessimo desiderato determinare l'effetto sulle variabili endogene di un

A. Arcelli

mutamento del parametro   a, supponendo cioè una trasposizione della
funzione del consumo mentre le spese governative rimanevano costanti,
avremo similmente ottenuto :

$$\left(\frac{\delta Y}{\delta \alpha}\right)^0 = \frac{\Delta_{31}}{\Delta} = \frac{1}{1-[b(1-z)+v]} \quad ; \quad \left(\frac{\delta Z}{\delta \alpha}\right)^0 = \frac{\Delta_{32}}{\Delta} = \frac{1-z}{1-[b(1-z)+v]} \quad ; \quad ecc.$$

I risultati sono identici a quelli ricavati supponendo una variazione della
variabile G : ciò è naturale perchè un mutamento del parametro a rappresenta una variazione   della domanda effettiva esattamente assimilabi-
le nei suoi effetti a quella prodotta da un mutamento di G (almeno
all'interno del nostro sistema interpretativo della realtà) .

Se conosciamo il campo di variabilità dei parametri strutturali,
senza tuttavia conoscere il loro esatto valore, non sempre possiamo determinare il segno delle derivate limitando le nostre informazioni al sistema statico di base .

Infatti nel nostro esempio particolare se non conosciamo esattamente il valore di b, di r   e di   v, è impossibile stabilire a
priori ad es. se ($\frac{\delta Y}{\delta G}$)^0 è alternativamente $\gtreqless$ 0 , poichè il
segno del denominatore è indeterminato.

L'economista a differenza dell'econometrico non fornisce stime
statistiche dei parametri strutturali : si accontenta di indicare il loro campo di variabilità, derivando le sue asserzioni da modelli" a latere"
relativi al comportamento degli operatori economici , che egli ha prece-
dentemente analizzato [10] In taluni casi la determinazione del campo di variabilità dei parametri permette di stabilire univocamente il segno dei
teoremi di statica comparata ; in altri, la soluzione si ottiene in virtù

---

[10] Su tale punto si veda Papandreou A. , L'Economia come scienza , op.
cit . , p. 153

M. Arcelli

del principio di corrispondenza che stabilisce un'intima relazione formale tra statica e dinamica. Di qui la necessità di ampliare il modello
di base mediante un'interpretazione aumentata e la formazione di modelli
dinamici corrispondenti al sistema di base. Questo è appunto il nostro
caso, che ci porterà a considerare due sistemi dinamici che incorporano le ipotesi del sistema statico e che sono, pertanto, ad esso corrispondenti . Un primo sistema sarà fondato su equazioni differenziali ; il secondo su equazioni alle differenze finite. Diversi saranno i procedimenti
nei due casi per fissare le condizioni di stabilità, ma entrambi permet-
teranno la determinazione univoca del segno delle derivate delle variabili endogene rispetto alle variabili esogene.

Prima di affrontare tale analisi dobbiamo tuttavia considerare in
una prospettiva allargata il problema della determinatezza delle proposizioni qualitative della statica comparata. Tale tema che costituisce attual-
mente materia di ampio dibattito tra gli economisti, può formularsi in
termini precisi con le parole di Lancaster [11] : "Si tratta di investigare
le condizioni in base alle quali le proposizioni economiche possono essere espresse in forma qualitativa, cioè in una forma in cui il segno al-
gebrico di qualche effetto è prevedibile basandosi esclusivamente sul segno dei parametri strutturali rilevanti del sistema".

Come è noto, Samuelson [12] aveva fornito una prima risposta
affermando che : "nei casi in cui i valori d'equilibrio delle variabili possono essere considerati come soluzioni di un estremo (massimo o mini-

---

[11] Si veda Lancaster K.J., The scope of qualitative economics , Review
of Economic Studies XXIX, febbr. 1962

[12] Si veda Samuelson P. , Foundations. of Economic Analysis, op. cit.,
p. 21.

M. Arcelli

mo) , è spesso possibile, indipendentemente dal numero di variabili considerate, determinare univocamente il comportamento qualitativo dei valori soluzione in connessione con mutamenti dei parametri". La risposta di Samuelson s'inquadra perfettamente nell'ambito del principio di corrispondenza che svela una relazione bidirezionale tra stabilità e posizioni di massimo di un corrispondente sistema statico. Lancaster riprende la questione per esaminare al di fuori delle informazioni attinte dalla dinamica , quali siano le condizioni in base alle quali le proposizioni economiche possono ricevere espressione qualitativa.

Egli affronta tale problema attraverso lo studio delle proprietà dei determinanti. Se la matrice dei segni dei coefficienti (nel nostro caso lo Jacobiano formato dai segni + - e 0, rispettivamente per coefficienti positivi, negativi e nulli) mediante l'intercambio di righe o l'intercambio di colonne o il rovesciamento dei segni di una riga o di una colonna, può essere posta in forma S [13], allora il segno dei teoremi di statica comparata è univocamente determinato. La forma S è caratterizzata dal fatto che tutti i segni sulla diagonale (o sulla quasi diagonale se la matrice non è quadrata) devono essere negativi ; tutti i segni sotto la diagonale devono essere nulli, mentre al di sopra della diagonale nessun segno può essere negativo e almeno qualcuno deve essere positivo.

Le conclusioni di Lancaster hanno però subito una revisione critica di vasta portata in uno scritto di Gorman [14]. Gorman mostra che esiste una classe ben più ampia di sistemi per cui risulta provabile la deter-

---

[13] Si veda su tale argomento una chiara esposizione di Papandreou A. , Introduction to macroeconomic models, Center of Plamning and Economic Research, Atene 1965, pp. 66-71

[14] Si veda Gorman W. M. More scope for qualitative economics, Review of Economic Studies, XXXI, genn. 1964 , pp. 65-68 .

M. Arcelli

minatezza qualitativa ; inoltre, come lo stesso Lancaster ha ammesso, le condizioni di Lancaster non sono applicabili a sistemi con un minor grado d'interdipendenza di quello da lui studiato. Gorman, seguendo un diverso indirizzo, ha pertanto proposto nuove condizioni di determinatezza qualitativa che includono quelle di Lancaster come caso particolare . In definitiva, il campo di applicazione delle condizioni di Lancaster si limita ai sistemi in cui siano predominanti le relazioni di inter di-pendenza [15] , mentre si è chiarito che un assottigliamento di tale rete facilita, in genere, la determinatezza qualitativa .

Nonostante i rilevanti progressi compiuti nel fissare le condizioni di determinatezza, il principio di corrispondenza occupa però ancora il ruolo principale nella soluzione dei problemi relativi al comportamento qualitativo delle variabili endogene. In questo senso la formulazione dei sistemi dinamici corrispondenti al nostro sistema statico di base, mantiene tutta la sua efficacia esemplificativa dei procedimenti di verifica dei teoremi di statica comparata, nell'ambito dell'economia qualitativa .

Possiamo quindi procedere considerando il seguente sistema, che costituisce una versione aumentata del sistema statico di base . La dinamizzazione viene attuata introducendo la seguente ipotesi [16]:

I rappresenta ora gli investimenti desiderati o ex-ante. Solo in condizioni di equilibrio tale grandezza è pari alla quantità risparmi-investimenti che si realizza effettivamente. Se vi è un improvviso aumento dei consumi, gli investimenti effettivi risultano inferiori agli inve-

---

[15] Si veda Lancaster K. , Partitionable Systems and Qualitative Economics, Review of Economic Studies , XXXI, genn. 1964 , p. 69 .

[16] Tale ipotesi viene utilizzata anche da Samuelson (si veda Foundations of Economic Analysis, op. cit. , p. 278) in un'applicazione del principio di corrispondenza al modello Keynesiano .

M. Arcelli

stimenti desiderati. Viceversa nel caso contrario . Nel primo caso vi sarà un impulso che tende ad accrescere il reddito; nel secondo caso si
verificherà l'effetto opposto. Possiamo supporre che il saggio di cambia-
mento del reddito sia proporzionale alla differenza tra investimenti desiderati e risparmi-investimenti realizzati. Il nostro sistema di base si tra-
sforma allora nel seguente sistema dinamico [17], contenente un'equazione
differenziale :

$$1) \quad \dot{Y} = I - \left[Y - C - G \right]$$
$$2) \quad Y - Z - T = 0$$
$$3) \quad C - a - bZ = 0$$
$$4) \quad I - u - vY = 0$$
$$5) \quad T - rY = 0$$

La prima equazione esprime appunto il saggio di variazione del
reddito in funzione della differenza tra investimenti programmati e risparmi-
mio : I rappresenta ora gli investimenti programmati. L'aumento dei consumi è determinato da una trasposizione della funzione del consumo rappre-
sentabile da un mutamento di a.

Le equazioni statiche sono così state sostituite da corrispondenti
equazioni che incorporano le ipotesi di base come casi particolari.

Da tale sistema si ricavano immediatamente le soluzioni che espri
mono il sentiero temporale delle variabili. Avremo pertanto.

---

[17] Il Prof. Lombardini sarebbe esitante ad attribuire la qualifica di modello
dinamico ad un tale modello che possiede soprattutto una fecondità euristica,
essendo invece limitata la sua rilevanza empirica. L'assenza dell'acceleratore limiterebbe il contenuto dinamico del modello. Pur accettando il rilie-
vo del Prof. Lombardini, preferiamo mantenere una terminologia oramai
accolta . L'introduzione dell'acceleratore, d'altro lato, avrebbe rotto la
corrispondenza con il modello statico e il conseguente isolamento avrebbe
lasciato insoluto il problema della determinatezza delle proposizioni qualitative.

M. Arcelli

$$Y(t) = \overset{\circ}{Y} + c_1 e^{\lambda t} \qquad \text{da cui} \qquad Y(t) = \overset{\circ}{Y} + (Y_0 - \overset{\circ}{Y}) e^{\lambda t}$$

$$C(t) = \overset{\circ}{C} + c_2 e^{\lambda t} \qquad \text{da cui} \qquad C(t) = \overset{\circ}{C} + (C_0 - \overset{\circ}{C}) e^{\lambda t}$$

$$I(t) = \overset{\circ}{I} + c_3 e^{\lambda t} \qquad \text{da cui} \qquad I(t) = \overset{\circ}{I} + (I_0 - \overset{\circ}{I}) e^{\lambda t}$$

eccetera

dove λ è la radice latente dell'equazione caratteristica data da :

$$
\Delta(\lambda) = \begin{vmatrix}
 & Y & Z & C & I & T \\
(-I-\lambda) & 0 & 1 & 1 & 0 \\
1 & -1 & 0 & 0 & -1 \\
0 & -b & 1 & 0 & 0 \\
-V & 0 & 0 & 1 & 0 \\
-r & 0 & 0 & 0 & 1
\end{vmatrix} = 0
$$

Sviluppando e risolvendo si ottiene $\lambda = \Delta = -I + b(1-z)+v$
a tale risultato si perviene anche attraverso una soluzione graduale
del sistema. Effettuando le sostituzioni nella prima equazione, essa infatti
può scriversi :

$$\overset{\bullet}{Y} = u + vY - \left[Y - a - b(1-r)Y - G\right] = 0 \qquad \text{da cui}$$

$$\overset{\bullet}{Y} = Y(v-1) + Yb(1-r) + G+a+u = 0 \qquad \text{ponendo} \quad G+a+u = A$$

$$\overset{\bullet}{Y} = Y\left\{-1 + \left[b(1-r) + v\right]\right\} + A \qquad \text{ma } -1 + \left[b(1-r) + v\right] = \Delta$$

$$Y(t) c_1 e^{\Delta t} - \frac{A}{\Delta} \quad \text{e poiché} \quad -\frac{A}{\Delta} = \overset{\circ}{Y} \quad Y(t) = \overset{\circ}{Y} + (Y_0 - \overset{\circ}{Y}) e^{\Delta t}$$

Da tale espressione possiamo ricavare poi I(t) ; C(t) ecc. Affinche

il sistema sia stabile, deve essere $\Delta < 0$ [18]. Infatti nel caso di e-
quazioni differenziali, le soluzioni dell'omogenea devono possedere parte
reale negativa. Da ciò discende la stabilità dell'equilibrio del sistema di
base : infatti da qualunque insieme di condizioni iniziali, tutte le variabili
tendono alle soluzioni d'equilibrio stazionarie al tendere del tempo all'infi-
nito.

Le condizioni di stabilità permettono di determinare il segno dei
teoremi di statica comparata, dedotti dal sistema di base. Infatti se Δ
è negativo, il denominatore delle espressioni delle incognite $(\frac{\delta X_k}{\delta G})^0$ è
positivo.

Conoscendo già che b, r e v sono positivi e minori di uno, il
segno di tutte le $(\frac{\delta X_k}{\delta G})^0$ sarà dunque positivo.

Costruiremo ora una versione aumentata del modello di base, fonda-
ta su equazioni alle differenze finite. Supponiamo che il consumo sia funzione del reddito disponibile del periodo precedente e che gli investimenti,
similmente, dipendano dal reddito del tempo t-1. Il sistema dinamico
sarà allora

1) $\quad Y_t - C_t - I_t - G_t = 0$

2) $\quad Y_t - Z_t - T_t = 0$

3) $\quad C_t - a - bZ_{t-1} = 0$

4) $\quad I_t - u - v Y_{t-1} = 0$

5) $\quad T_t - rY_t = 0$

sostituendo nella prima equazione

---

[18] Ciò implica che il sistema sarà stabile solo se la somma della propensione al consumo rispetto al reddito nazionale b(1-r) e della propensione
all'investimento **v**, è inferiore all'unità.

$$Y_t - a - b(1-r) \quad Y_{t-1} - u \, v Y_{t-1} - G_t = 0 \qquad \text{da cui}$$

$$Y_t - \left[b(1-r) + v \right] \quad Y_{t-1} - a - u - G_t = 0 \qquad \text{e ponendo } a+u+G = A$$

$$Y_t = \left[b(1-z) + v \right] \quad Y_{t-I} + A \quad \text{e risolvendo}$$

$$Y_t = c_1 \left[b(1-r)+v) \right]^t + \frac{A}{1 - \left[b(1-r)+v \right]} \quad \text{e poichè} \quad \frac{A}{1 - \left[b(1-r)+v \right]} = \overset{o}{Y}$$

$$Y_t = (Y_o - \overset{\bullet}{Y}) \quad \left[b(1-r)+v \right]^t + \overset{o}{Y}$$

Similmente possiamo ottenere C_t, I_t ecc.

Affinchè il sistema sia stabile deve essere

$$- 1 < \left[b(1-r) + v \right] < 1 \quad \text{e poichè l'espressione tra parentesi è}$$
positiva, basterà $0 < \left[b(1-r) + v \right] < 1$, cioè ancora $\triangle < 0$.

La condizione di stabilità permettono anche in questo caso di risolvere il problema della determinatezza qualitativa delle proposizioni economiche.

Appare quindi evidente che nel processo di verifica di una semplice proposizione economica si è spesso costretti ad ampliare ed eventualmente a riformulare il modello di base per rendere possibile il confronto con la realtà. Spesso è necessario introdurre considerazioni probabilistiche che completano le versioni aumentate del modello, trasformando il sistema in struttura stocastica, oltre che dinamica.

Dalla nostra esemplificazione risulta pure che esistono diverse versioni aumentate di un identico modello di base; pertanto, se le ipotesi che l'economista intende verificare sono incorporate nel modello di base, egli non respingerà le sue assunsioni salvo che tutte le possibili versioni aumentate del modello non siano state respinte dal confronto con l'evidenza empirica. In questo senso assume significato il concetto di verifica indiretta e globale delle ipotesi del modello di base e si comprende che esso "gode di un alto grado di isolamento dalla realtà"[19]

---

[19] Si veda Papandreou A. , L'economia come scienza, op. cit. , p. 160

CENTRO INTERNAZIONALE MATEMATICO ESTIVO

(C. I. M. E.)

Harold W. KUHN

SOME REMARKS ON GAMES OF FAIR DIVISION

Corso tenuto   a Villa Falconieri (Frascati) dal 22 al 30 agosto 1966

SOME REMARKS ON GAMES OF FAIR DIVISION

by

Harold W. Kuhn **✴**

There are at least two reasons that make research in the subject of games of fair division seem needed. First, at the base of welfare economics there is the question of which political and social systems ensure the "equitable" or "fair" distribution of economic goods . Secondly, the theoretical literature on such systems is almost non-existent, consisting in the main of isolated rules for solving special and simple situations, often unaccompanied by any rigorous proof. We shall not change this situation very much but propose to reopen some old questions, giving them new formulations and adding a minor bit of precision here and there.

Any rigorous discussion of the problem should begin with some definitions. Fortunately, for the purposes of this lecture we need not provide these in any detail. We shall work primarily from examples and have formal definitions for an appendix of the printed version.

All discussion of " fair- division" begin with the classical method for dividing a "continuous" object (such as a cake) into fair shares for two individuals. The solution of this problem has been known for a long time : one player divides the cake into two pieces and the other player chooses one of the pieces for himself. If we wish to formalize this, we might let S denote the set which is the cake, let $N = \{1, 2\}$ denote the players and let $\mathscr{P} = \{\{S_1, S_2\}\}$ denote the class of physically realizable partitions of the cake into two pieces (S_1 and S_2, where $S_1 \cap S_2 = \emptyset$ and $S_1 \cup S_2 = S$) . Each player should have a rule for

✴ Princeton University . This research was supported by the National Science Foundation ; a somewhat different version is to be published in a volume of essays on mathematical economics honoring 0 . Morgernstern to be published by the Princeton University Press (1966) , edited by M. Shubik .

H. W. Kuhn

deciding whether the physically realizable piece T is "acceptable" or "fair" for him. We may represent this by a function f_i for $i \in N$, where $f_i(T) = 1$ if player i considers the piece $T \in \{S_1, S_2\} \in \mathcal{P}$ fair and $f_i(T) = 0$. otherwise. Then the classical rule of divide-and -choose can be presented as the extensive form of a game as follows :

Move 1. The players are assigned the roles of Divider and
Chooser through the toss of a fair coin.

Move 2. The Divider selects $P = \{S_1, S_2\} \in \mathcal{P}$

Move 3. The Chooser selects $X = (x_{ij})$ where i, j = 0 or 1
and $\sum_i x_{ij} = \sum_j x_{ij} = 1$ for i, j = 1, 2.

Payoff : Player i receives share S_j if and only if $x_{ij} = 1$.

This scheme is fair in the sense that each player can assure himself a fair share through the use of an appropriate strategy if we make the following assumptions :

1) no matter how the cake is divided into two pieces, each player will find at least one of the pieces to be acceptable ;

2) either player is able to divide the cake so that both pieces are acceptable to him.

(Note that we have introduced the additional feature of an initial chance move to ensure that the outcome does not depend on the labeling of the players ; traditionally, there is the social convention that the older player divides the cake and the younger player chooses) .

Steinhaus [8] has proposed a generalization of this scheme for three players. We shall present his method as the extensive form of a game, using the notation introduced above.

H. W. Kuhn

Move 1. The players are assigned the roles of one Divider and two Choosers by means of a chance device yielding equal probabilities to the tree possible assignments.

Move 2. The Divider selects $P = \{S_1, S_2, S_3\} \in \mathcal{P}$.

Move 3. Each Chooser i announces which of the shares S_j are acceptable to him, i.e., the S_j for which $f_i(S_j) = 1$.

Move 4. If a share can be assigned to each Chooser that is acceptable to him, then this is done and the remaining share is given to the Divider. Otherwise, some share is unacceptable to both Choosers. This share is assigned to the Divider and the Choosers divide the remaining shares according to the two-player fair division scheme.

In order to insure that this scheme is fair, we need only assume that :

1) no matter how the cake is divided into three pieces, each player will find at least one of the pieces acceptable to him ;

2) each of the players is able to divide the cake so that each of the three pieces is acceptable to him ;

3) if a piece found unacceptable by both of the Choosers is assigned to the Divider then the remaining parts are considered by the Choosers as a fair amount to divide among themselves (i.e., (1) and (2) hold for the resulting two-player fair division scheme) .

Steinhaus [3] reports a "divide-and-choose" scheme for n players due to B. Knaster and S. Banach. Their solution is :

" The partners being ranged 1, 2, 3, . . . , n , cuts from the cake an arbitrary part. 2 has now the right, but is not obliged, to diminish the slice cut off. Whatever he does, 3 has the right (without obligation) to diminish still the already diminished (or not dimi-

H. W. Kuhn

shed) slice, and so on up to n. The rule obliges the "last diminisher" to take as his part the slice he was the last to touch . This partner thus disposed of, the remaining n - 1 persons start the same game with the remainder of the cake. After the number of participants has been reduced to two, they apply the classical rule [one divides while the other chooses] for halving the remainder " .

This scheme, which seems to be equivalent to the method in which an Umpire moves a knife across the cake until some player calls "stop " and receives the resulting piece, is very ingenious. However, it is not a direct generalization of the schemes discussed above. Our first result is a scheme which extends the two - and three-player schemes in a direct manner. We begin with an example.

Let us assume that there is a cake to be divided among seven players. Let a player , say 1, be designated Divider by an equiprobable chance device . We shall assume that player 1 can divide the cake into pieces $S_1, S_2, \ldots, S_7$ so that $f(S_j) = 1$ for $j = 1, \ldots, 7$. Define the 7 by 7 matrix $A = (a_{ij})$ by $a_{ij} = f_i(S_j)$ for $i, j = 1, \ldots, 7$. For the sake of an example, this might look as follows :

	S_1	S_2	S_3	S_4	S_5	S_6	S_7
P_1	1	1	1	1	1	1	1
P_2	1	0	0	0	0	0	0
$P_{,3}$	0	0	1	0	0	1	0
P_4	0	0	0	0	0	1	1
P_5	0	1	0	0	1	0	0
P_6	1	0	0	0	0	0	0
P_7	0	0	0	0	0	1	1

H.W. Kuhn

I assert that this problem can be reduced to a fair division problem among four players as follows : Assign the piece S_5 to P_1, the piece S_3 to P_3, and the piece S_2 to P_5 . These pieces (S_2, S_3, S_5) are acceptable to each of these players and have been declared unacceptable by the players (P_2, P_4, P_6, P_7) who have not been assigned pieces. Therefore, we shall assume that P_2, P_4, P_6, and P_7 each consider the remainder of the cake or fair share for the four of them collectively.

In effect , we have made a definition of a "fair reduction" of an n player cake problem as the reduction of the division problem to a smaller set of m players by assigning n-m parts of S to players who find them acceptable, while none of these n-m parts are acceptable to the m players remaining . We then generalize (1) and (2) above to assume that (in a fair reduction to a set of m players and a set T of cake remaining) :

1) if T is divided into m parts , then each player will find at least one of the parts acceptable ;

2) each player is able to divide T into m parts, each of which is acceptable to him.

To complete our example to a theory we need a combinatorial lemma which asserts that a fair reduction is always possible.

Definition. Given an n by n matrix $A = (a_{ij})$ with entries 0 or 1, a (partial) assignment is a set $\left\{(i_1, j_1), \ldots, (i_r, j_r)\right\}$ where $I = \left\{i_1, \ldots, i_r\right\}$ and $J = \left\{j_1, \ldots, j_r\right\}$ each contain $r \geq 1$ distinct indices, and $a_{ij} = 1$ for $i \in I$ and $j \in J$. An assignment is complete if r = n.

Lemma. Let $A = (a_{ij})$ be an n by n matrix with entries 0 or 1 such that $a_{1j} = 1$ for $j = 1, \ldots, n$ and $\sum_j a_{ij} \geq 1$ for $i = 2, \ldots, n$. Then there exists an assignment such that $a_{ij} = 0$ for $i \notin I$ and $j \in J$.

Proof. Either there is a complete assignment or, by the theorem of Frobenius-König, there exists an s by t rectangle of zeros with s+ t= = n + 1. Choose s to be a maximum. If s = n - 1 then

$$A = \begin{bmatrix} 1 & 1 & \ldots & 1 \\ 0 & a_{22} & \ldots & a_{2n} \\ \cdot & \cdot & & \cdot \\ \cdot & \cdot & & \cdot \\ \cdot & \cdot & & \cdot \\ 0 & a_{n2} & \ldots & a_{nn} \end{bmatrix}$$

by reordering the rows and columns and   (1, 1) is an assignment. Other-wise, we may reorder the rows and columns of A   so that :

(note that M has at least one row and three columns since   s + t = and $s \leq n$ - 2 imply t - 2 $\geq$ 1.) Since s is a maximum , there is no k by t rectangle of zeros in M with (s + k) + 1 = n+1. That is, there is no k by 1 rectangle of zeros in M with k + 1 = (n+1) - s = t. Now consider the (t - 1) by t matrix

H. W. Kuhn

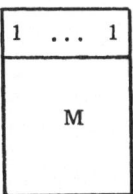

and delete the first column . In this (t-1) by (t - 1) matrix , there is no k by 1 rectangle of zeros with $k + 1 = (t - 1) + 1$. Hence there exists a complete assignment for this matrix which is an assignment for A satisfying the condition of the Lemma that $a_{ij} = 0$ for $i \notin I$ and $j \in J$.

Q. E. D.

Knaster has also suggested (cf. Steinhaus [3]) a method of division applicable to situations in which S is indivisible. i.e. assumption (2) does not hold. To illustrate his scheme, suppose that three heirs must share four objects among themselves. They are first asked to reveal the monetary value of each object to an Umpire. Denote the value of object i to heir j by v_{ij} . An example is tabulated below :

		Heir	
	1	2	3
Object 1	$1,000	5,000	1,000
2	3,000	1,000	4,000
3	3,000	2,000	1,000
4	5,000	4,000	6,000

If we assume that the monetary value of sets of objects is additive, then heir j believes his fair share is $\frac{1}{3} \sum_{i} v_{ij}$ is $4,000 for each j in our example Knaster's rule assigns each object to an heir who values it most highly and then defines sidepayments (summing to zero) among the heirs so as to insure an equal surplus over the fair share for each . The amount of this surplus is exactly

H. W. Kuhn

$$\frac{1}{n} \left(\sum_i (\max_j \ v_{ij}) - \sum_j \frac{1}{n} \sum_i v_{ij} \right).$$

Applying this rule to the example :

$$\text{Heir} \begin{cases} 1 \\ 2 \text{ receives} \\ 3 \text{ object (s)} \end{cases} \begin{cases} 1 \\ 2 \\ 2,4 \end{cases} \text{ valued at} \begin{cases} 3,000 \text{\textcrossed{\$}} \\ 5,000 \\ 10,000 \end{cases} \text{and a side payment of} \begin{cases} 3,000 \text{\textcrossed{\$}} \\ 1,000 \\ -4,000 \end{cases}$$

Note that since the fair share of each is $ 4,000, the outcome gives each
heir the same surplus of $ 2,000 . It is a simple and straighforward
matter to verify that this rule provides a non-negative surplus no matter
what the valuations are . However, we propose to show how Knaster's rule can be "discovered" through linear programming and simultaneously
demonstrate the properties claimed for it.

To establish some notation, let v_{ij} be the valuation of object
i by heir j, where   i = 1 ,..., m and j = 1,...,n . Then $v_j = \frac{1}{n} \sum_i v_{ij}$
is the fair share of heir j. As before X = (x_{ij}) will be an assignment
matrix defined by x_{ij} = 1 if object i is assigned to heir j and
x_{ij} = 0 otherwise . Finally, let y_j denote the sidepayment to heir j.

Primal Program. Find $(x_{ij}), (y_j)$, and z so as to maximize
z subject   to $x_{ij} \geq 0$ and

$$\sum_i v_{ij} x_{ij} + y_j - v_j \geq z \qquad (j = 1, \ldots, n)$$

$$\sum_j x_{ij} = 1 \qquad (i = 1, \ldots, m)$$

$$\sum_j y_j = 0 .$$

The interpretation of the program should be clear. We seek to
maximize the surplus z   over the fair share which can be given to
all players , using side payments y_j which sum to zero. A straightforward

H. W. Kuhn

application of the theory of duality yields :

<u>Dual program</u>. Find $(s_j), (r_j)$ and t so as to   minimize $-\sum_j v_j s_j +$
$+ \sum_i r_i$ subject to $s_j \geq 0$ and

$$\sum_j s_j = 1$$

$$- v_{ij} s_j + r_i \geq 0 \quad (i = 1, \ldots, m \text{ and } j = 1, \ldots, n)$$

$$- s_j + t = 0 \qquad (j = 1, \ldots, n) .$$

Since all $s_j = t, nt = 1$ and $s_j = t = \frac{t}{n}$ for all j and we may rewrite the dual :

Find (r_j) so as to minimize $\sum_i r_i - \frac{1}{n} \sum_j v_j$ subject to

$$r_i \geq \frac{1}{n} v_{ij} \quad (i = 1, \ldots, m \text{ and } j = 1, \ldots, n) .$$

The solution of this program is immediate. Set $n\bar{r}_i = \max_j v_{ij}$
Then

$$n^2 \bar{r}_i \geq \sum_j v_{ij}$$

$$n^2 \sum_i \bar{r}_i \geq \sum_{i,j} v_{ij} = n \sum_j v_j$$

$$\sum_i \bar{r}_i \geq \frac{1}{n} \sum_j v_j$$

which implies that the optimal value of the dual program is nonnegative. This completes an existence proof for the primal, ignoring for the moment the question of whether the x_{ij} take on integral values. Howe-ver, we may solve the primal explicitly.

For each i, set $\bar{x}_{ij} = 1$ for one j for which $v_{ij} = \max_j v_{ij}$. The primal program then becomes : Find (y_j) and z so as to maximize z subject to $\sum_i v_{ij} \bar{x}_{ij} + y_j - v_j \geq z \ (j = 1, \ldots, n)$

$$\sum_j y_j = 0$$

H. W. Kuhn

Summing the first set of constraints

$$\sum_{i,j} v_{ij} \bar{x}_{ij} - \sum_j v_j \geq nz \ .$$

If we set

$$\bar{z} = \frac{1}{n} \ (\sum_i (\max_j v_{ij}) - \sum_j v_j)$$

this gives equality in the sum and hence equality in all of the inequality summands. Note that $\bar{z} \geq 0$, by our result for the dual program. Moreover,

$$\bar{y}_j = v_j - \sum_i v_{ij} \bar{x}_{ij} + \frac{1}{n} \ (\sum_i (\max_j v_{ij}) - \sum_j v_j)$$

clearly sum to zero. The three terms in $\bar{y}_j$ are, respectively, the fair share of j, the value of the objects assigned to j, and the surplus. Thus the Knaster procedure is completerly verified.

The-feeling of euphoria which may have been created by the two results proved above may be dispelled by the following two examples.

Example 1 . Suppose S consists of six pieces of cake, one with a cherry and the others plain, to be divided by P_1 and P_2 No piece of cake can be divided further. The acceptable sets are defined by giving additive utility functions. Namely, both P_1 and P_2 value a plain piece of cake at 1 utile, P_1 places a value of - 1 utile on the piece with the cherry, and P_2 values the piece with the cherry at 3 utiles. If we denote the resulting utility functions by u_1 and u_2, respectively,

$$\mathscr{F}_1 = \{T \mid u_1 (T) \geq 2\}$$
$$\mathscr{F}_2 = \{T \mid u_2 (T) \geq 4\} \ .$$

Clearly, if A divides and B chooses, A will select $S_1 = \{2 \text{ pieces of plain cake}\}$ and $S_2 = \{3 \text{ pieces of plain cake, 1 piece with the cherry}\}$, and the utility outcome will be $u_1 = 2$, $a_2 = 6$.

H. W. Kuhn

On the other hand, if B divides and A chooses, the obvious division yields the utility outcome $u_1 = 4$, $u_2 = 4$. If either divider has knowledge of the chooser's utility function, it is easy to verify that the utility outcome is $u_1 = 3$, $u_2 = 5$. Simple as our example is, it illustrates phenomena which are not covered by our previous theory namely the "chooser's advantage" and the moderating effect of the knowledge of the chooser's preferences.

Example 2. Let us modify our example of three heirs and the four objects by assuming that Heir 2 "cheats" by announcing false values to the Umpire. The following table gives these false values :

		Heir		
		1	2	3
	1	1,000	1,001	1,000
Object	2	3,000	3,999	4,000
	3	3,000	2,999	1,000
	4	5,000	6,001	6,000

Application of Knaster's rule yelds

$$
\text{Heir}
\begin{cases}
1 \\
2 \\
3
\end{cases}
\text{receives object (s)}
\begin{cases}
3 \\
1, 4 \\
2
\end{cases}
\text{valued at}
\begin{cases}
\$ 3,000 \\
7,002 \\
4,000
\end{cases}
\text{and a side payment of}
\begin{cases}
\$ 1,445 \\
- 1,890 \\
445
\end{cases}
$$

Note that the true value to Heir 2 is $ 9,000 - $ 1,890 = $ 7,110 compared to the $ 6,000 which he received originally. The surplus value to the other players has been reduced to $ 445.

The numbers in this example have been chosen only to exhibit the advantages that can accrue to a player who falsely portrays his own valuations with a knowledge of the other player's true valuations. It points up a clear need for an analysis of the strategic opportunities of this situation.

H. W. Kuhn

APPENDIX AND BIBLIOGRAPHY

Here we shall present some tentative definitions. They may seem overly elabotate for the application given them above; however, they have been designed to provide a framework for a quite general development of the subject. .

<u>Definition</u> 1. A fair division problem is defined by $(S, N, \mathcal{P}, \mathcal{F})$ where S is a set to be divided among the players $N = \{1, \ldots, n\}$ and $\mathcal{P}$ is a family of partitions $P = \{S_1, \ldots, S_n\}$ of S into n-subsets. The family $\mathcal{F}$ consists of classes $\mathcal{F}_i$ of subsets T of S for $i \in N$. Player i considers any set $T \in \mathcal{F}_i$ to be acceptable or fair for his share.

<u>Definition</u> 2. A legal division is an assignment of the sets S_j of a partition $P \in \mathcal{P}$ to the players $i \in N$. Such an assignment may be denoted by $X = (x_{ij})$ where $x_{ij} = 1$ if S_j is given to player i and $x_{ij} = 0$ otherwise. If S_{ji} is assigned to player i and $S_{ji} \in \mathcal{F}_i$ for $i = 1, \ldots, n$, then the legal division is called a fair division.

<u>Definition</u> 3. A fair division scheme is the extensive form of a game in which all outcomes are legal divisions and in which each player can assure himself a fair share through the use of an appropriate strategy.

The definition of a fair reduction of a division game given above may be formalized as follows :

Let $T \subset N$ and let $M \subset N$ contain m players. The restriction of $\mathcal{P}$ to T and M will be denoted by $\mathcal{P}(T, M)$ and consists of all partitions $\{S_1, \ldots, S_m\}$ of T such that there exists $\{S_1, \ldots, S_m, S_{m+1}, \ldots, S_n\} \in \mathcal{P}$. (Note that $\mathcal{P}(S, N) = \mathcal{P}$). Suppose $P = \{S_1, \ldots, S_m, S_{m+1}, \ldots, S_n\} \in \mathcal{P}$ is such that $\bigcup_{j=1}^{m} S_j = T$ and $S_j \notin \mathcal{F}_i$ for $i \in M = \{i_1, \ldots, i_m\}$ and $j = m + 1, \ldots, n$. Then $\mathcal{P}(T, M)$ is called a fair restriction of $\mathcal{P}$ to T and M . (note that

H. W. Kuhn

$\mathcal{P}$ (S, N) is a fair restriction) .

We shall assume, for all fair restrictions $\mathcal{P}$ (T, M) of $\mathcal{P}$ to T and M, that :

(1) for all $i \in M$ and all $P \in \mathcal{P}(T, M)$, $P \cap \mathcal{F}_i \neq \phi$;

(2) for all $i \in M$, there exists a $P_i \in \mathcal{P}(T, M)$ such that $P_i \subset \mathcal{F}_i$.

The fair division scheme which is shown possible by the Lemma is then formalized as follows :

Fair division scheme . Let a Player, say 1, be designated Divider by an equiprobable chance device. Then player 1 selects $P = \{S_1, \ldots, S_n\} \in \mathcal{P}$ such that $S_j \in \mathcal{F}_1$ for all j . (This is possible by (2)). Define $A = (a_{ij})$ by $a_{ij} = 1$ if $S_j \in \mathcal{F}_i$ and $a_{ij} = 0$ otherwise. The matrix A satisfies the hypotheses of the Lemma. (Note that $\sum_j a_{ij} \geq$ for i = 2, ..., n by (1) .)

Hence there exists an assignment $\{(i_1, j_1), \ldots, (i_r, j_r)\}$ with $I = \{i_1, \ldots, i_r\}$, $J = \{j_1, \ldots, j_r\}$ with $S_j \in \mathcal{F}_i$ for i $\in$ I and j $\in$ J . Assign S_j to Player i for i $\in$ I and j $\in$ J. Then , if $T = \bigcup_{j \notin J} S_j$ and $M = \{i \ / \ i \notin I\}$, the restriction of $\mathcal{P}$ to T and M is a fair restriction and the process may be repeated. Since r $\geq$ 1, at least one player i receives a part $S_j \in \mathcal{F}_i$ at each iteration, and hence the scheme terminates. Whenever S_j is assigned to Player i we have $S_j \in \mathcal{F}_i$ and hence the scheme is fair.

This note was begun with a disclaimer that no general theory of fair division would be presented. The truth of this should be obvious to the reader who has penetrated this far. However, some of the needs are clear and the field seems open to the diligent researcher. To aid anyone who would heed this call, a bibliography of sources known to the author is appended .

H. W. Kuhn

[1] Knaster, B. and Steinhaus, H., Ann. de la Soc. Polonaise de Math.,
    19, 228-31, 1946 .

[2] Steinhaus, H., "The Problem of Fair Division", Econometrica,
    16, 101-104 , 1948 .

[3] _____ ," Sur la division pragmatique", Econometrica, 17 (supple-
    ment) , 315-319 , 1949 .

[4] G. Th. Guilbaud , "Les problèmes de partage, " Économie Appliqué, I,
    93-137, 1952 .

[5] Luce, R. D. and Raiffa, H., Games and Decisions, New York, 1957 .

[6] Flood, M., "Some Experimental Games, " Management Science, 5,
    5-26, 1958 .

[7] Shubik, M. , Strategy and Market Structure ; Competition , Oligopoly,
    and the Theory of Games, New York, 1959 .

[8] Steinhaus, H., Mathematical Snapshots, 2nd edition, New York, 1960 .

[9] Dubins, L. E. and Spanier, E. H., "How to Cut a Cake Fairly, " Ame-
    rican Math Monthly, 68, 1-17, 1961 .

[10] Singer, E, "Extension of the Classical Rule of 'Divide and Choose'",
    Southern Economic Journal, 38, 1962 .

CENTRO INTERNAZIONALE MATEMATICO ESTIVO

(C. I. M. E.)

Harold W. KUHN

ON TWO THEOREMS IN INTERNATIONAL TRADE

Corso tenuto a Villa Falconieri (Frascati) dal 22 al 30 agosto 1966

ON TWO THEOREMS IN INTERNATIONAL TRADE

by Harold W. Kuhn*

INTRODUCTION. In a recent survey of the theory of international trade [2], Chipman has devoted a major portion of the
third part, entitled "Modern Theory" to two theorems, the Samuelson
factor price equalization theorem and the Stolper-Samuelson theorem
on the effects of tariffs on real wages. The history of these two
theorems is a complicated one and the literature that they have
generated is voluminous. In this lecture, I shall focus my attention
on some mathematical aspects which have been of interest to me; the
limited time permits only a rapid outline of the economic background.
Fortunately, the Chipman article will provide an easy entry to the
subject for anyone who is interested in pursuing it further. Furthermore, the excellent bibliography to be found in Chipman's survey
will allow us to cite articles by the year and refer the reader to
his listing.

The Samuelson theorem (1948, 1949) has as its source the
revision of classical international trade theory undertaken by
Heckscher (1919) and Ohlin (1933). Ohlin, whose work was known in

*This research was done while the author was National Science Foun-
dation Senior Postdoctoral Fellow and Visiting Professor at the
University of Rome.

the English-reading world before that of Heckscher due to a thirty
year delay in the translation of the 1919 paper, had asserted that
there is a tendency toward factor price equalization as a result
of free trade.   However, he qualified his statement in a number
of directions, even to the point of claiming that equalization
would never be complete. Heckscher, on the other hand, made a
clear claim for the equality of factor prices and it was this
assertion that Samuelson attempted to prove. I shall attempt to
sketch the final form that this proof has assumed and indicate the
nature of the new mathematics which it has stimulated.

The background of the Stolper-Samuelson theorem is stated
by Chipman as follows (underlining added): "While Heckscher had
stated the factor price equalization theorem in terms of complete
equalization, Ohlin (1933) stated it only in terms of tendencies.
Stolper and Samuelson (1941) proved a very strong theorem concerning
these tendencies, namely that trade would lower the price of the
"scarce" factor expressed in terms of any commodity. Thus
Samuelson's 1948 and 1949 papers represented the evolution of what
was an outgrowth of Ohlin's more cautious statement, back full
circle to Heckscher's original position. So great has been the
attention given to the conditions for full equalization, that
the problem originally posed by Stolper and Samuelson has remained
relatively underdeveloped."   Chipman goes on to sketch a proof of
the Stolper-Samuelson theorem. With minor modifications, this
proof is valid but only for the two good - two factor case; it is

unusual that this restriction has not been more widely noted. The lecture concludes with some modest attempts to extend the result to more goods and factors.

Turning first to the Factor Price Equalization Theorem, let n goods be produced in amounts $y_1, \ldots, y_j, \ldots, y_n$ using m factors of production, indexed by $i = 1, \ldots, m$. The production functions

$$y_j = f_j(x_1, \ldots, x_m)$$

are assumed to be concave homogeneous functions of the first order defined for all nonnegative inputs. Let factor prices $w_1, \ldots, w_m$ be given and define

$$C_j(y_j; w_1, \ldots, w_m) = \min\{\Sigma_i w_i x_i \mid y_j = f_j(x_1, \ldots, x_m)\} \ .$$

As Shephard [5] first proved, this minimum cost function factors as follows:

$$C_j(y_j; w_1, \ldots, w_m) = y_j \cdot g_j(w_1, \ldots, w_m) \ ;$$

the minimum unit cost functions can be shown to be continuous, homogeneous of the first order, and concave.

If we let x_{ij} be the quantity of factor i used in producing goods j , then a competitive equilibrium for this production economy may be defined as quantities of goods y_j and factors x_{ij} and prices of goods p_j and factors w_i such that

$$p_j \frac{\partial f_j}{\partial x_{ij}} \leq w_i \ ,$$

where if strict inequality holds then $x_{ij} = 0$, and

$$\Sigma_i w_i x_{ij} \geqq p_j \; ,$$

where if strict inequality holds then $y_j = 0$.

It is known [6] that the first set of conditions are necessary and sufficient conditions for each set of factor inputs $(x_{1j}, \ldots, x_{mj})$ to achieve the minimum unit cost $g_j(w_1, \ldots, w_m)$ for the j-th goods at factor prices $(w_1, \ldots, w_m)$. Thus, the existence of an equilibrium may be formulated in the following terms, which apply equally well to nondifferentiable production functions (arising, say, from a linear programming setting). Given the production functions f_j and positive prices p_j for the goods, there exist a set of nonnegative factor inputs $(x_{1j}, \ldots, x_{mj})$ such that

$$g_j(w_1, \ldots, w_m) = w_1 x_{1j} + \ldots + w_m x_{mj} \geqq p_j \; ,$$

where if strict inequality holds then $y_j = 0$.

Up to now, we have dealt with only one country; however, since we assume identical production functions in each country, we have the same vector inequality

$$g(w) \geqq p$$

as the condition of equilibrium in each country. Therefore, as Chipman has pointed out, "factor price equalization will take place if (1) all n commodities are produced in each country (so that $y_j > 0$ and $g_j(w) = p_j$ for all j , and (2) the function g is globally invertible." For then the goods prices p_j will determine factor prices w_i , which will be the same in all countries.

Restating (2), we must show that, given p_j , there are some factor endowments for which $g(w) = p$ has a solution and then we must show that the solution is unique. The existence of a solution was handled rigorously for the first time by Kuhn in 1959 [7], although Chipman has quite properly criticized the restrictive assumptions under which the theorem was proved. The proof is so short that we shall reproduce it here. The questionable hypothesis under which it was proved is

INTENSITY HYPOTHESIS: Suppose $m = n$ and that there is an indexing of goods and factors (associating with each goods the factor that is used intensively in its production) such that $w_i = 0$ implies

$$g_i(w_1,\ldots,w_m)/p_i < g_k(w_1,\ldots,w_m)/p_k$$

for some k . (Informally, this asserts that, if the factor that is used intensively in goods i is free, then goods i is not the most expensive to produce.)

PROOF. For each nonzero set of factor prices, define $h(w_1,\ldots,w_m) = \max_i g_i(w_1,\ldots,w_m)/p_i$. Then set

$$w_i' = w_i + h(w_1,\ldots,w_m) - g_i(w_1,\ldots,w_m)/p_i .$$

This defines a continuous mapping of nonzero factor prices into themselves. By the Brouwer fixed-point theorem, there exist factor prices $\bar{w}_1,\ldots,\bar{w}_m$ and a constant $c > 0$ such that

$$c\bar{w}_i = \bar{w}_i + h(\bar{w}) - g_i(\bar{w})/p_i \quad \text{for all } i .$$

Choose k such that $g_k(\overline{w})/p_k = h(\overline{w})$. For this k , $c\overline{w}_k = \overline{w}_k$ and hence either $c = 1$ or $\overline{w}_k = 0$. The latter possibility is ruled out by the Intensity Hypothesis. Hence

$$g_i(\overline{w}) = h(\overline{w})p_i \quad \text{for all} \quad i \ .$$

By the homogeneity of g_i ,

$$g_i(\overline{w}/h(\overline{w})) = p_i \quad \text{for all} \quad i \ ,$$

and the existence of a solution is proved.

The question of uniqueness is delicate; sufficient conditions with a natural economic interpretation seem difficult to find. Samuelson's first attempt, which involved an incorrect application of the implicit function theorem, has led Nikaidô and Gale [8] to develop new results on the global univalence of mappings. The principle theorem is that a sufficient condition for the global univalence of a differential mapping $g(w) = p$, where both w and p are m-dimensional is that the Jacobian matrix $[\partial g_i/\partial w_j]$ have all positive principle minors.

Now turning to the Stolper-Samuelson theorem, attempts to
extend it to many commodities and factors from the two commodity
and two factor case leads to an interesting and difficult technical
problem. We shall follow the discussion of Chipman (see [2],
pp. 37-39) in stating the problem. As he has noted, "the argument
of the Stolper-Samuelson theorem, as has been made clear by Bhagwati
[4], breaks into two distinct steps: (1) trade lowers the relative
price of the commodity that employs the "scarce" factor relatively
intensively; (2) the fall in this relative price brings about a
fall in the price of the "scarce" factor relative to all commodity
prices." We shall consider only the second step of the argument.

For the case of an equal number of commodities and factors,
Chipman (see [2] or [3]) has isolated what is needed to validate (2)
in a particularly elegant style. Namely, assuming that each factor
has been associated with the commodity in which it is used intensively, then we must show $\partial \log w_i / \partial \log p_i > 1$, where p_i and
w_i are the prices of the associated commodity i and factor i .
That is, we must show that the elasticity of the i-th factor price
relative to the i-th commodity price must exceed unity. These
elasticities may be shown to be the diagonal elements of the matrix
inverse to

$$A = (a_{ij}) = (\frac{1}{p_i} x_{ij} w_j)$$

where x_{ij} is the quantity used of factor j in making one unit of
commodity i at minimum cost when factor prices are $(w_1, \ldots, w_n)$.

Clearly A is a stochastic matrix; that is, all $a_{ij} \geqq 0$ and $\Sigma_j a_{ij} = 1$ for $i = 1,\ldots,n$. Thus, the second step of the Stolper-Samuelson theorem is reduced to showing that the inverse of A has all of its diagonal elements greater than one.

At this point, Chipman has made a minor slip. We quote (see [2], p. 38): "For the case $n = 2$ it is always true that the inverse of a stochastic matrix (when it exists) has its diagonal elements either greater than one or less than zero; so by appropriate permutations of rows and columns (that is, by suitable association of commodities with their "intensive" factors -- indeed, this provides us with a definition of "intensity"), these diagonal elements will always exceed unity." Clearly, this cannot be true since $A = \begin{pmatrix} 1 & 0 \\ 0 & 1 \end{pmatrix}$ is a stochastic matrix. A correct statement is provided by the following theorem.

THEOREM 1. Let $A = \begin{pmatrix} a_{11} & a_{12} \\ a_{21} & a_{22} \end{pmatrix}$ be a stochastic matrix with positive determinant. Let $A^{-1} = \begin{pmatrix} b_{11} & b_{12} \\ b_{21} & b_{22} \end{pmatrix}$. Then $b_{11} \geqq 1$ and $b_{22} \geqq 1$. Furthermore if $a_{12} > 0$ then $b_{11} > 1$ and if $a_{21} > 0$ then $b_{22} > 1$.

PROOF. (Before proceeding to the proof itself, note that there is no restriction involved in the assumption that the determinant is positive; if the inverse exists then the determinant is nonzero and, with the possible permutation of the rows, can be assumed positive.)   By direct calculation,

$$b_{11} = \frac{a_{22}}{a_{11}a_{22} - a_{12}a_{21}} \quad \text{and} \quad b_{12} = \frac{-a_{12}}{a_{11}a_{22} - a_{12}a_{21}}$$

and hence

$$b_{11} + b_{12} = \frac{a_{22} - a_{12}}{a_{11}a_{22} - a_{12}a_{21}} = \frac{a_{22} - a_{12}}{a_{22}(1-a_{12}) - a_{12}(1-a_{22})} = 1 .$$

Hence $b_{11} \geq 1$, and if $a_{12} > 0$ then $b_{11} > 1$. The proof for b_{22} is exactly analogous.

It should be noted that the assumption that A has a positive determinant is, in effect, an "intensity hypothesis." Namely, $a_{11}a_{22} - a_{12}a_{21} > 0$ implies

$$\frac{1}{p_1} x_{11}w_1 \frac{1}{p_2} x_{22}w_2 - \frac{1}{p_1} x_{12}w_2 \frac{1}{p_2} x_{21}w_1 > 0$$

and hence, if all of the prices are positive,

$$x_{11}x_{22} - x_{12}x_{21} > 0 .$$

To insure $b_{11} > 1$ and $b_{22} > 1$, we must have all of $x_{ij} > 0$ and hence this may be written:

$$\frac{x_{11}}{x_{12}} > \frac{x_{21}}{x_{22}} .$$

This says that relatively more of factor 1 is used in making one unit of commodity 1 than in making one unit of factor 2.

Before proceeding to the case of three commodities and three factors we shall establish several general properties which might otherwise appear to be peculiar to the case just considered.

PROPOSITION 1. If A has each of its row sums equal to one and if $B = A^{-1}$, then each of the rows of B has sum equal to one.

PROOF. Let $A = (a_{ij})$ and $A^{-1} = B = (b_{ij})$. Then
$$\Sigma_j b_{ij} = \Sigma_j b_{ij} (\Sigma_k a_{jk}) = \Sigma_k \Sigma_j b_{ij} a_{jk} = 1.$$

PROPOSITION 2. If A has each of its row sums equal to one and $a_{kk} a_{ij} > a_{kj} a_{ik}$ for all k and all i, $j \neq k$, then
$$a_{kk} > a_{ik} \text{ for all } k \text{ and all } i \neq k.$$

PROOF. (Note that the conclusion of this proposition, namely, that the diagonal elements of A be larger than any other element in the same column, is the condition that Chipman has announced [3] as being sufficient for the case of $n = 3$ but not for $n \geq 4$.) Assuming $a_{kk} a_{ij} > a_{kj} a_{ik}$ all k, all i, $j \neq k$ we have
$$\Sigma_{j \neq k} a_{kk} a_{ij} > \Sigma_{j \neq k} a_{kj} a_{ik}.$$

That is,
$$a_{kk}(1-a_{ik}) > a_{ik}(1-a_{kk})$$

and hence
$$a_{kk} > a_{ik} \text{ all } k, \text{ all } i \neq k.$$

THEOREM 2. (Chipman) Let $A = (a_{ij})$ be a 3 by 3 stochastic matrix and let $a_{kk} > a_{ik}$ for all k and all $i \neq k$. Then A is nonsingular. Let $A^{-1} = B = (b_{ij})$. Then $b_{kk} \geq 1$ for all j and, if $a_{kj} > 0$ for some $j \neq k$ then $b_{kk} > 1$.

PROOF. If A can be shown to be nonsingular then

$$(b_{12} + b_{13}) \det A = (a_{13}a_{32} - a_{12}a_{33}) + (a_{12}a_{23} - a_{13}a_{22})$$

$$\leqq (a_{13}a_{22} - a_{12}a_{33}) + (a_{12}a_{33} - a_{13}a_{22}) = 0$$

where strict inequality holds if $a_{12} > 0$ or $a_{12} > 0$. If $\det A$ can be shown to be positive then

$$b_{12} + b_{13} \leqq 0$$

where strict inequality holds if $a_{13} > 0$ or $a_{12} > 0$. Since, by Proposition 1, $b_{11} + b_{12} + b_{13} = 1$, we have

$$b_{11} \geqq 1$$

where strict inequality holds if $a_{12} > 0$ or $a_{13} > 0$. Thus the truth of the theorem depends on the

LEMMA. $\det A > 0$.

PROOF. We shall first prove A is nonsingular. If not, $XA = 0$ for some $X \neq 0$. Then $\Sigma_i x_i a_{ij} = 0$ and hence $0 = \Sigma_j \Sigma_i x_i a_{ij} = \Sigma_i x_i \Sigma_j a_{ij} = \Sigma_i x_i$. Since $X \neq 0$, we may assume $x_1 = 1$, $x_2 \leqq 0$, $x_3 \leqq 0$ and $x_2 + x_3 = -1$, that is, that the first row of A is a convex combination of the other rows of A. (Note that this may necessitate the simultaneous reordering of the rows and columns of A which will not change the column domination of the diagonal.) However, this implies

$$a_{11} = (-x_2)a_{21} + (-x_3)a_{31} < a_{11}(-x_2-x_3) = a_{11},$$

which is a contradiction.

However, the set of stochastic matrices for which $a_{kk} > a_{ik}$ $(i \neq k)$ is a convex set. The function det A is continuous on this set and det I = 1 for the matrix I which lies in the set. Since det A is never zero we must have det A > 0 throughout. This completes the proof of the lemma and of the theorem.

The case of more than 3 goods is still entirely open. First note that Chipman's condition (that the diagonal elements of A be larger than any other element in the same column) does not even imply that A is nonsingular. The following example shows this:

$$
\begin{pmatrix}
\frac{1}{3} & \frac{1}{6} & \frac{1}{4} & \frac{1}{4} \\
\frac{1}{6} & \frac{1}{3} & \frac{1}{4} & \frac{1}{4} \\
\frac{1}{4} & \frac{1}{4} & \frac{1}{3} & \frac{1}{6} \\
\frac{1}{4} & \frac{1}{4} & \frac{1}{6} & \frac{1}{3}
\end{pmatrix}
$$

However, no example is known of a 4 by 4 stochastic matrix satisfying the intensity hypothesis of Proposition 2 that is singular. Hence there is some (vague) hope that some intensity hypothesis may be strong enough to prove the Stolper-Samuelson theorem for four factors and four commodities.

BIBLIOGRAPHY

[1]   STOLPER, WOLFGANG F., and SAMUELSON, PAUL A.: "Protection
      and Real Wages," Review of Economic Studies, 9 (November,
      1941), 51-73. Reprinted in Readings in the Theory of Inter-
      national Trade. Philadelphia: The Blakiston Company 1949,
      pp. 333-357.

[2]   CHIPMAN, JOHN S.: "A Survey of the Theory of International
      Trade: Part 3, The Modern Theory," Econometrica, 34
      (January, 1966), 18-76.

[3]   _____: "Factor Price Equalization and the Stolper-
      (abstract) Samuelson Theorem," Econometrica, 32 (October,
      1964), 682-683.

[4]   BHAGWATI, JAGDISH: "Protection, Real Wages and Real Income,"
      Economic Journal, 69 (December, 1959), 733-749.

[5]   SHEPHARD, RONALD W.: Cost and Production Functions.
      Princeton: Princeton University Press, 1953.

[6]   KUHN, HAROLD W., and TUCKER, ALBERT W.: "Nonlinear Programming,"
      Proceedings of the Second Berkeley Symposium on Mathematical
      Statistics and Probability. Berkeley: University of Cali-
      fornia Press, 1951, pp. 481-492.

[7]   KUHN, HAROLD W.: "Factor Endowments and Factor Prices:
      Mathematical Appendix," Economica, N.S., 26 (May 1959),
      142-144.

[8]   GALE, DAVID, and NIKAIDÔ, HUKUKANE: "The Jacobian Matrix
      and Global Univalence of Mappings," Mathematischen
      Annalen, 159, Heft 2 (1965), 81-93.

CENTRO INTERNAZIONALE MATEMATICO ESTIVO

(C. I. M. E.)

A. PAPANDRE OU

THEORY CONSTRUCTION AND EMPIRICAL MEANING IN ECONOMICS

Corso tenuto a Villa Falconieri (Frascati) dal 22. al 30 agosto 1966

THEORY CONSTRUCTION AND EMPIRICAL MEANING IN ECONOMICS

by

A. PAPANDREOU

I. On the impossibility of non taxonomic theory in economics.

We start with the proposition :

today in economics we do not have or "cannot" have a theory that is universal, which in other words is not taxonomic.

My first argument will be : that theory in economics today is of necessity taxonomic.

When I use the word "theory" I mean theory that has testability. I am not referring to an analytical system.

I make a distinction of course between analytic mathematical and logical propositions and theory, in the sense that the latter involves data and makes statements about data and can be tested by reference to data . It is empirical and synthetic so to say .

I argue that today it is impossible to have universal synthetic propositions in economics.

The question is : what is a formal representation of a theory?

It is natural for an econimist to start speaking about a system of equations that represent the theory. For example: The Walrasian system, the most elegant and the most complete representation of an economic theory.

But, what do we mean by a formal representation of a theory?. This is the question.

a. We posit a set of equations in implicit form.

(1.0)
$$F_i(x) = 0, \quad i = 1, 2, \ldots, I$$

where x is an ordered n-tuple

b. They characterize a structure:

---

[1] The brackets on the left of the text denote the parts that have added to the manuscript from notes taken during the lectures

A. Papandreou

(1.1)
$$\Phi = \left\{ F_1, F_2, \ldots, F_i, \ldots, F_n \right\}$$

where F_i is the set of the n-typles x such that $F_i(x) = 0$

provided : $\bigcap_i F_i \neq \emptyset$ is

satisfied.

That is to say, that the intersection of the subsets F_i is not an empty space.

One has to look at this structure as a purely formal concept and see how, from this formal concept, one moves on to the empirical world. It becomes necessary to find some operational rules in order to test the theory .

c. Let there be an empirical rule r for carrying out observations, and let A_r be the subset of all possible observation acts which result from the application of rule r.

We posit the mapping

(1.2)
$$r : A_r \longrightarrow X \qquad x = r(a)$$

where both x and a are ordered n-tuples

X is the cartesian space .

"x" is the image under rule "r" of "a" , "a" is an observation act. In this way wa have interpreted the theory .

The next question here is : what form does theory take ?

d. A theory may be cast in the following form : $\bigwedge_{a \in A_r} \left[r(a) \in \bigcap_i \right.$

For all "a", "r(a)" is an element of the intersection (the solution set) .

Such a theory is <u>universal</u> and <u>non analytic</u>. Its propositions are refutable, but are also sure to be refuted.

A. Papandreou

e. This holds even if the propositions are qualitative i.e. if they refer to "model" in common usage of the term (classes of structures), and therefore to the property of the $\bigcap_i F_i$ rather than to the $\bigcap_i F_i{}'$ s themselves.

g. We conclude : <u>universal, non analytic statements in economics must take on a taxonomic character, lest they be refuted outright</u> .

If we have taxonomic propositions, two questions arise : what form do they take? and can they be tested or not?

h. We introduce the taxonomic character : let W be the set of states of the world, Ω its power set, and ω a subset of W (an element of Ω) .

Let there be an empirical rule s for carrying out observation acts, and let V_s be the subset of all possible observation acts, which result from the application of rule s .

We posit the mapping :

$$(1.4) \qquad s : V_s \longrightarrow \Omega \ , \ \omega = s\,(v)$$

ω is the image under rule s of v .

We may now write with explicit reference to social space :

$$(1.5) \qquad \bigwedge_{v \,\in\, V_s} \bigwedge_{a \,\in\, A_{rs}} \left[s(v) = \omega^* \longrightarrow r_s \ (a) \in \bigcap_i F_i \right]$$

where $r \neq s$ and ω^* is some fixed social space .

This means that, if under rule s we find that the state is ω^* , then all observations must satisfy the solutions set .

It will be noted that in (1.5) we have written r_s in lieu of r.

A. Papandreou

This is intended to convey a dependence of r on s which guarantees that the two types of observation acts are carried out in the same space-time segment.

j. Thus our theory has taken on a taxonomic character. It relates to generically described (through rule s) social space ω^* .

Now the problem is : can we test this theory?

The answer is yes, if we have specified the rules r and s .

From the two, the rule s is the most difficult to specify .

The conclusion, that I arrive at this moment, is that the universal taxonomic propositions can be found applicable or not.

- o -

II - On the impossibility of refuting taxonomic universal propositions.

a. Given the state of our knowledge , rule s must remain uninterpreted. Thus the expression

$$s(v) = \omega^*$$

must remain uninterpreted.

b. In lieu of (1.5) we should write :

(2.0)
$$\bigwedge_{v \, \epsilon \, V_s} \bigwedge_{a \, \epsilon \, A_{rs}} \left[(S(v) = \omega^*_{un.}) \rightarrow r_s(a) \, \epsilon \bigcap_i F_i \right]$$

c. For the theory to be confirmed it is sufficient that

$$r_s(a) \, \epsilon \bigcap_i F_i$$

be confirmed.

d. For the theory to be <u>refuted,</u> however, we are obliged to show that

$$r_s(a) \in \bigcap_i F_i \quad \text{is} \quad \text{FALSE}$$

while

$$s(v) = \omega^* \quad \text{is} \quad \text{TRUE} .$$

This we cannot do; as long as rule s remains unintepreted.

e. We conclude : <u>Universal, taxonomic statements, for the pre-</u><u>sent can only be confirmed (or found inapplicable) by reference to em-</u><u>pirical data</u> .

We choose to call (2.0) a <u>model rather than a theory</u> .

(This is the sense in which the term "model" will be used here).

-. o -

III - <u>On the possibility of deriving refutable descriptive statements from</u><u>models.</u>

a. Transform (2.0) into a <u>descriptive</u> statement by writing

(3.0) $$\bigwedge_{a \in A_r} \left[a \ P \ H^* \longrightarrow r(a) \in \bigcap_i F_i \right]$$

where a P H^* conveys the notion that observation acts, a , take place in H^*, H^* being a given historical individual (a given segment of space time).

b. (3.0) is capable, in principle, of being refuted, but it lacks <u>universality</u> which is characteristic of theory .

c. If (3.0) has been <u>confirmed</u> for H^*, then we say that it is

A. Papandreou

an explanation   of H^{*} ; while if it has not been tested , we say that it is a prediction about H^{*} .

- o -

IV - On a general formulation of the taxonomic approach.

a. We define the mapping

$$(4.0) \qquad\qquad g : \Omega \to \Gamma , \qquad g(\omega) = \gamma$$

where $\gamma \in \Gamma$ and γ is a unit vector of J dimensions.

b. We   define a vector of structures

$$(4.1) \qquad (\phi^1, \phi^2, \ldots, \quad \phi^j, \ldots \phi^J)$$

and , correspondingly, the vector of solution sets

$$(4.2) \qquad (\bigcap_{i_1} F^1_{i_1} , \bigcap_{i_2} F^2_{i_2} , \ldots \bigcap_{i_j} F^j_{i_j} , \ldots \bigcap_{i_J} F^J_{i_J})$$

or more simply

$$(4.3) \qquad\qquad F^{*} = (F^{*1}, F^{*2}, \ldots, F^{*j}, \ldots, F^{*J})$$

c. It is clear that

$$(4.4) \qquad\qquad F^{*} g(\omega) = F^{*} g (s (v)) = F^{*} h (v) = F^{*j}$$

d. A generalized taxonomic theory (for an interpreted s) takes the form

$$(4.5) \qquad \bigwedge_{v \in V_s} \bigwedge_{a \in A_{r_s}} \left[r_s (a) \in F^{*} h (v) \right]$$

A. Papandreou

This means that for all observation acts   v and for all observations a, observations a, under rule r,   will   be in the cross product
of the vector solution sets of structures.

v   does two things: not only does it identify the state ω , but
fixes in the same time the solution set.

- o -

V - On the possibility of constructing a model corresponding to a gene-ralized taxonomic theory .

a. Let rule s   be interpreted, but be inadequate in selecting
the subset of states of the world, which through g . pick   the apprio-
priate solution sets.

Thus we may continue writing

$$s : V_s \longrightarrow \Omega , \quad \omega = s(v)$$

but cannot write

$$g : \Omega \qquad , \quad g(\omega) =$$

This means that the knowledge of ω leads   you to select
one   solution set γ .

The   v   does two things. It given us ω and ω gives us the
solution   set. But we do not know   anything about g,   because we do not
know anything about   the rule   s.

b. Let us assume further, however,   that   if we posit that

$$a P H^{*}$$

$$v P H^{*}$$

a rule g^{*} can   be formulated   such   that

- 128 -

A. Papandreou

(5.0)
$$g^* : \Omega \to \Gamma, \ g^*(\omega) = \gamma$$

c. Thus we may write

(5.1)
$$\bigwedge_{v \in V_s} \bigwedge_{a \in A_{rs}} \left[v \, P \, H^* \text{and} \ a \, P \, H^* \to r_s(a) \in F^* \, h^*(v) \right]$$

This means that :

If v and a are observations of specific social space in a specific time, then we can have a theory applicable to that period and that space.

d. It is capable of being refuted, but lacks universality. It may be an explanation, if confirmed, or a prediction, if untested.

c. It is much richer than (3.0)

- o -

VI - On classes of empirically meaningful statements in economics.

A. For non-generalized taxonomic models or descriptions of types (2.0) and (3.0) we have:

1. Statements concerning the properties of $\bigcap_i F_i$.

2. Statements relating to the causal ordering of the variables in a structure. (Also the direction of change of endogenous variables as a result of a change in a given direction of data and exogenous variables) .

B. For generalized taxonomic models or descriptions of types (4.5) or (5.1) , we have:

1. Statements involving comparisons of two or more distinct space- time segments (relating to features of type A_s or A_r) .

2. Statements involving the impact of organizational change on

A. Papandreou

any of the properties of some structure in a given space-time segment.

II   - MODELS OF DECISION -

Coming on to planning, there is a certain procedure in which one can thing so far as the scientific activity of planning is concerned in terms of some order in which one does things ; one, may be, should think of the planner as a chooser.

There arise of course all the questions for him that arise for the houshold or for the firm.

The standard questions of decision theory, emerge here, and one has to know what is his bundle of commodities, what is the constraint in his case, to formulate the decision of the planner and what are the criteria of choice .

In other words, we have to transfer to him the theory of choice, and see him as an actor.

We start out with the theory of choice .

I - A model of rational choice.

a. We limit the discussion to the case of certainty, namely the case where the outcome of each choice is expected by the actor with probability of 1.

This theory can be extended easily to the case of uncertainty .

b. The actor (choice-agent) is confronted at any one time with a choice set   X .

The alternatives contained in X are denoted by x and are n-tuples characterizing the relevant features of the state of the world.

A. Papandreou

$$x \quad \epsilon \quad X$$
$$x \quad = (x_1, x_2, \ldots, x_N)$$

where x_n is a quantity of some commodity or a dimension of a situation, etc. .

c. The fundamental axiom is that, for the actor in question and at a given moment of time (in a given context), the set X is <u>completely ordered</u> by a relation " $\gtrsim$ " meaning "not inferior to". This assumption of complete ordering is equivalent to the following pair of assumptions.

1. For any x', x'' : $x' \gtrsim x''$ or $x'' \gtrsim x'$

2. For any x', x'', x''' :

If $x' \gtrsim x''$ and $x'' \gtrsim x'''$ then $x' \gtrsim x'''$

They are respectively the axioms of <u>comparability</u> and of <u>transitivity</u> .

d. If $x' \gtrsim x''$ and $x'' \gtrsim x'$

then we write

$$x' \sim x'' \quad or \quad x'' \sim x'$$

(the symbol $\sim$ means indifference)

If $x' \gtrsim x''$ and $x'' \not\gtrsim x'$

then we write

$$x' > x''$$

(the symbol $>$ means preference)

Thus we define respectively <u>indifference</u> and <u>preference</u>.

So indifference and preference are definitions that are derived from the more general relation which contains them, namely "not inferior to" .

A. Papandreou

This is the foundation stone of the theory of rational choice.

I would like to say that it is terribly demending in any concrete choice situation, and between the two the most demending is the comparability axiom, namely that there is always a response when somebody is present with a pair-wise choice (either this or that) . For example , if you are given to choose where would you rather live between two geographic locations.

When you have no experience of either of the two, it would be very hard for you to answer. So experience is the key for the comparability axiom.

Choice, when you are not familiar with the objects of choice, is very painful indeed.

On the other hand the transitivity axiom is not too demanding, because, after all, it goes to the foundation of one's consistency. This is the first part of our theory. Choice has, of course, to be completed.

What we have already said defines the value system, the tastes of the chooser. Now we have to define his environment.

e. Not all elements are accessible to the actor. Thus we introduce the feasible set

$$\overset{o}{X} \subset X$$

f. Finally , we define __equilibrium__ for the actor.

The actor is in equilibrium, if, and only if, he has chosen some $\overset{\wedge}{\underset{\bullet}{x}} \in \overset{o}{X}$ so that for all $\overset{o}{x} \in \overset{o}{X}$, $\overset{\wedge}{\underset{o}{x}} \succsim \overset{o}{x}$

Namely he will pick one element of the feasible set, in such a fashion that it is 'not inferior to' any of the other elements of the feasible

A. Papandreou

set $\overset{o}{X}$.

So there are really three headings here : the tastes, the constraints and a rule, and this is the rule that defines the equilibrium. It is a maximization act and of course this can be translated into utility theory under certain conditions.

g. An interesting result obtains if we add certain mathematical restrictions which will not be examined here .

There exists an infinite class U of mappings, any mapping in which, say   u , has the form

$$u : X \longrightarrow R$$

where   R   is   the class of reals . These mappings have the following property :

For any u ϵ U , for any   x', x" ϵ X :

$$u(x') \geqslant u(x")$$

if , and only if,   x' $\geqslant$ x"

Thus, u   is   a   <u>utility index</u> which <u>represents</u>   the preference system of the actor, and preserves the order to   x   by $\geqslant$.

Thus it becomes possible to   show   that the equilibrium alternative $\overset{\wedge}{\overset{o}{x}}$ has the property that,   for   any   u ϵ U

$$u(\overset{\wedge}{\overset{o}{x}}) = \underset{\underset{x \epsilon X}{o}}{ma x} \; u (\overset{o}{x})$$

Under the circumstances,   it is permissible to argue that   an actor is in equilibrium if , and only if, he has <u>maximized</u> some utility index, but in this particular form theory is not very useful because it requires too much for us if we wanted to be actors.

A. Papandreou

We then lower our sights to a theory which is not as demanding, a theory which we may call of limited rationality.

- o -

II- A model of limited rationality in choice.

a. This model is obtained by _relaxing_ the requirements of rationa-lity , by introducing psychological considerations, and by making the approach more _operational_ .

b. The actor _perceives_ a set of _feasible_ choice alternatives.

The feasible set is _evoked_ (transferred from _passive_ to _active_ memory, where memory is viewed as a set generated by _learning_) by stimuli.

There is a _dynamic interaction_ between the _evoked_ set and the _evoking stimuli._

c. Here we must make the crucial distinction between _choice_ and _outcome of choice._

Let $\overset{o}{Y}$ be the _perceived feasible set of choices_ available to the actor, where

$$\overset{o}{y} \in \overset{o}{Y} \text{ and } \overset{o}{y} = (\overset{o}{y}_1, \overset{o}{y}_2, \dots \overset{o}{y}_n) .$$

Let X be the set of outcomes, where

$$x \in X \text{ and } x = (x_1, x_2, \dots x_N)$$

Furthermore, $\overset{*}{X}$ is the power set of X namely , the set of all possible subsets of X.

d. We now introduce the concept of _belief_ or _expectation_ or

A. Papandreou

<u>information</u> via the mapping

$$\mathbf{f} \; : \quad \overset{o}{Y} \longrightarrow \overset{*}{X}$$

in other words, what we are seeing here, is that for each choice made, there will be a possible set of outcomes.

Thus, to each $\overset{o}{y} \in \overset{o}{Y}$ there corresponds under f a set of outcomes, $X_{\overset{\bullet}{y}}$

$$X_{\overset{\bullet}{y}} = \mathbf{f}(\overset{o}{y})$$

where $X_{\overset{\bullet}{y}} \in \overset{*}{X}$

and $\underset{\overset{o}{y} \in Y}{\overset{o}{\bigcup}} X_{\overset{\bullet}{y}} = \overset{o}{X} \subset X$;

$f(\overset{o}{y})$ is the expectation (or the information) function. It is the actor's expectation of what his act is going to do to the environment.

e. If the $X_{\overset{\bullet}{y}}$ become singleton sets, then we have the case of certainty. We can easily introduce risk and uncertainty in the model, but we shall not attempt it here.

f. A preference relation in a very simple form is introduced setting 1 stand for <u>satisfactory</u> or <u>acceptable</u> and 0 for <u>un-satisfactory</u> or <u>unacceptable</u>, we have a <u>valuation</u> mapping

$$w \; : \quad \overset{o}{X} \longrightarrow \{0, 1\}$$

The mapping w characterizes the <u>aspiration</u> level of the actor . This level is dependent upon the experience of the actor : the degree of success in obtaining acceptable outcomes.

g. Finally, in lieu of a definition of equilibrium of the actor , we introduce a <u>search process</u> which details in a <u>dynamic</u> fashion the manner

A. Papandreou

in which the actor attempts to arrive at a <u>state of rest</u>.

I really think that the planner, as I see him in a real choice situation, is best described in this fashion.

I said that we have a learning process and I think it best describes his behavior.

Now we should transfer this model, to the language of the variables of the planner, and here I should select a Tinbergen's type model to link up what I have just done.

- o -

III - The planner's decision.

a. We will employ the limited rationality model of choice in a very schematic fashion and assume that the belief function leads to singleton sets.

b. Which is the choice set $\overset{o}{Y}$ confronting the planner ?

I think that the most useful way of looking at this problem, is looking at the time path of the economy, (x_c). The objects are really time paths of the economy.

This x_t is a vector of time paths of the economy (consumption, investment, imports, exports). We define the set of time paths :

$$x_t \in X$$

the choice is obviously of picking some time paths (some behavior of the system overtime).

But what is feasible ? What is feasible is not given to you in

A. Papandreou

any sense or fashion, but in any-thing else except your own theory and your own model.

Now let us assume that you are in a very lucky position of having such a theory.

We recognize <u>exogenous variables</u> within a structure ϕ subject to the control of the planner .

$$\xi \;=\; (\xi_1, \xi_2, \ldots \xi_j)$$

and <u>parameters of institutional</u> (structural) <u>change</u>

$$\pi = (\xi_1, \xi_2, \ldots \xi_I)$$

Let $\overset{o}{y} \;=\; (\xi, \pi)$

$\overset{o}{y} \;=\; \overset{o}{Y}$

$\overset{o}{y}$ is the choice : the picking of the value of an exogenous variable or a parameter. These are not the outcomes.

c. We posit a system of dynamic relations characterized by a set of equations, whose solution takes the form

$$\overset{o}{x}_t = g\,(\,t,\, \alpha\,,\, \delta\,,\, \overset{o}{y}\,)$$

where $\overset{o}{x}$ is an n-tuple of endogenous variables and $\overset{o}{x}_t$ is the n-tuple of their <u>sequences</u> , and where α is the n-tuple of <u>initial condi</u>-tion**s**, and δ the n-tuple of <u>data,</u> variables beyond the control of the planner which appear, mathematically speaking, on additional initial condi-tions .t stands for time .

d. $\overset{o}{X}_t$ is the set of perceived feasible n-tuples of sequences .

It is generated by $\overset{o}{x}_t = g(t, \alpha, \delta, \overset{o}{y})$

for all admiss ble values of δ , $\overset{o}{y}$.

A. Papandreou

e. Using the valuation mapping w, we select some subset of $\overset{o}{X}_t$, and by introducing some additional process for selection, we end up with an acceptable sequence $\overset{\wedge}{\overset{o}{x}}_t$.

f. This leads to a choice of $\overset{o}{y}$, given g.

g. Finally, we introduce a learning process on a national scale. Let

$$g = f \, (\, \tilde{x}_t - \overset{o}{x}_t \,)$$

where $\tilde{x}_t$ is the <u>observed</u> as against the <u>expected</u> sequence $\overset{o}{x}_t$.

f should have the property that as $t \longrightarrow \infty$ the sequence tends to the zero sequence.

h. We may link up the planner's decision to the methodological considerations with which we started the discussion.

The $\tilde{x}$'s correspond to the a's under rule r . $\overset{*}{H}$ defines the space-time segment given by the planner's horizon. Finally, g characterizes the $\underset{i}{\bigcap} F_i'$s .

- o -

IV - <u>The task of the planning bureau, in practice.</u>

a. The <u>first task</u> is <u>SCANNING</u> for feasible <u>TIME-PATHS,</u> so that the <u>FEASIBLE SUBSET</u> may be formulated.

b. We have already argued that the <u>FEASIBLE SUBSET</u> must be defined - ideally in terms of a dynamic model whose solution is of the form

$$x_t = g \, (t, \alpha \, , \, \delta \, , \, \varepsilon \, , \, \pi \,)$$

where ε are the <u>INSTRUMENT</u> variables and π the <u>INSTITUTIONAL</u>

A. Papandreou

CHANGE variables.

c. We have admitted that ideally x_t should be replaced by a subjective probability distribution over a set of time-paths, so that to eache, say, ε or π corresponds under g a probability distribution rather than some given time-path .

d. It is understood, of course, that the model employed constitutes a prediction on the part of the Bureau of planning for the social space involved. It corresponds, therefore, to the case

$$\bigwedge_{v \,\in\, V_s} \quad \bigwedge_{a \,\in\, A_{rs}} \left[v \ P \ \overset{*}{H} \ \text{ and } \ a \ P \ \overset{*}{H} \longrightarrow r_s(a) \in \overset{*}{F} \ h(v) \right]$$

(Description of the individual space-time segment in question).

e. It is safe to assume that in almost all instances the lack of data and of relevant, useful hypotheses precludes this general formulation, or restricts it to highly aggregative data which are not sufficiently useful for planning decisions.

With the development of a statistical service oriented to processing the kind of information which is needed by the bureau, and through a learning process we may converge toward the development of the required model.

f. In practice, especially the early stages, there exists a process for scanning feasible time-paths which is simpler and workable, though not intellectually satisfactory.

1 Project the time-paths on the basis of an econometric model based on the data of recent experience.

2. Consider marginal departures of the planned time-paths from

A. Papandreou

the projected time-paths .

These departures are associated with something that you already know . You have had some communications from the political authority, some indications of the general targets which they would like.

3. Search into their feasibility by inquiries into technological and institutional questions, one by one .

4. Test the consistency of a set of simultaneous deviations by the use of a variety of models both in the real and in the monetary sectors.

5. Arrive at a list of limited alternatives.

g. One major qualification is now in order. The POLITICAL AUTORITY to which the results must be submitted has probably already set some limitations on the range of the alternatives to be considered in the feasible set .

1. Concerning the π 's

2. Concerning the scope of the $\mathfrak{t}$'s .

h. The second task of the Planning Bureau is the submission of the feasible set to the Political Authority, so it may apply its valuation rule and select the OPTIMAL SUBSET .

i. Actually, the Planning Bureau cannot expect the Political Autority to order completely the alternatives. At most, the Political Authority will be prepared to make some relevant PAIR-WISE CHOICES, which will suggest PARTIAL ORDERING of alternatives;

j. The pairwise choices submitted to the Political Authority must be of a critical nature.

They must pose to it the basic choices open to the country in

A. Papandreou

question for the period in question.

    k. Actually, if the Bureau of planning had <u>initially</u> demanded
of the Political Authority a commitment on the general tergets it wishes
to pursue, the task of selecting feasible alternatives would be much sim-
plified.

    1. A feed-back mechanism leading from <u>incremental information</u>
to model revision <u>(annual rolling plans)</u> is an essential feature of the
planning operation from a scientific point of view. LEARNING ON A NATIO-
NAL SCALE.

    m. Thus far we have examined planning as a scientific process.
We have not concerned ourselves with its importance or relevance as
a <u>POLITICAL DOCUMENT</u>, as a <u>CONTRAT SOCIAL</u>.

    For it has far-reaching significance on :

1. The ratio of consumption to national income ;
2. The personal distribution of wealth ;
3. The functional distribution of income ;
4. The regional distribution of economic activity ;
5. The extent to which non-domestic resouces will be used and
    the conditions under which they will be accepted ;
6. The role of the state in the process, and the freedom reser-
    ved for the citizen .

    I would like to add one more thing about planning. Usually in a
short term plan (5 years plan) , where the horizon is limited, many of the
government's activities (such as expenditures in education or certain kinds
of investments) seem to make no sense for there is no pay-off for the
5 years period in question .

    The ideal way of thinking is to ask yourself what will be the ima-

ge of the country you have for the period of 15 - 20 years. If you make
alternative assumptions about these things, they may have implications
on what   investment activities the government may undertake today .

- o -

V - Some remarks about types of planning.

a. There is an intimate relationship between the general targets
set by the POLITICAL AUTHORITY   and the restrictions it imposes
on the SCOPE OF THE INSTRUMENT VARIABLES   or on the permissible
set of INSTITUTIONAL CHANGES .

b. A key   question which emerges is this : How centralized or
decentralized   is the planning process envisaged by the political autority?

c. Decentralized decision-making in an economy fundamentally means
that first, information processing takes place in a decentralized fashion,
being carried out by the basic behavior agents, households and firms. Second, each basic behavior agent behaves according to a general rule or
strategy whether self-imposed or not. Thus, decentralized decision-making
is compatible with planning, as is the case when the rule of behavior is
imposed by the planning authority (Lange, Lerner) .

d. I am prepared to argue that the effectiveness of a plan varies
directly with degree of centralization, while its efficiency varies directly with the degree of decentralization.

e. The size-structure of the horizon of the unit in connection
with exernalities or non-decomposability is a critical question concerning
the efficiency of a plan as well as the whole process of economic
development.

f. We introduce the concept of dual planning which best describes

A. Papandreou

planning in action.

      <u>Sector A</u> is either planned in a centralized fashion or in a decentralized fashion.

      <u>Sector B,</u> in contrast, is guided toward the targets by the central authority by affecting the <u>environment of decision</u> of the units.

      Here again we can construct an index of <u>DEGREE OF DECENTRA-LIZATION</u> by taking account of the relative significance of the two sectors and the type of planning employed within sector A.

A. Papandreou

PLANNING AS A SOCIAL PROCESS

I made a distinction between centralized and decentralized planning.

Decentralized planning is the case where the information processing is made by the agents of the economy, namely the households and the firms, but the rules of action of these agents are not determined by their own choice but rather are imposed upon them by some rule , the strategy being communicated to them by some central authority. So you have decentralization and at the same time you have planning. At the other end you have centralized planning in the extreme form , of which there is only one decision agent, that is the central bureau, which determines the targets for all members of the economy.

What you find in reality is a mixture of these two techniques, a mix of planning and non planning.

Since this is a more realistic way of approaching an economic system I would like to suggest very simply how one can do this.

The first thing to do is to divide the economy to two sectors : the sector which is planned and the sector which is unplanned.
Let us consider that A is the planned sector.

Here again we can make a distinction and consider the centralized planned sector and the decentralized planned sector.

The larger the sector A is against the sector B (unplanned sector) the more planned the economy is.

I come now to the question that prof. Lombardini raised two days ago, namely the question of the issue of looking at planning not as an

A. Papandreou

insider but as an outsider . That   is to say : do we have a theory of
planning as a   social process and not as a scientific process ?

But the social process of planning goes to the heart of politics
and sociology.

Now I   shall   give you some loose observations and loose gene-
ralizations, because a theory of this type is not available.

Here the language of game theory would be useful.

If the plan is not a serious social decision then nobody follows
it and it is just an exercise.

But if the plan is really meant for action then it becomes the
focus of great social conflict. Then one can see the plan as a "contrat
social" among the vested interests of the country, regional, national and other
interests. This necessitates that we talk a little more about power, and
specifically political power. We can identify this power with the exercise
of influence by anybody over the various branches of the formal political
machinery of a state. What   the constitution tells us is only what is the
formal distribution of the authority, to make certain kinds of decision. But
the process of influence over those who do make the decisions, has to be
studied by political scientists. But because of the invisibility of this parallel authority it is very difficult   for political scientists to study it.

And one of its objectives is to remain invisible, because is this
invisibility that allows it to function.

In the classical concept of a democracy, each citizen exercises
equal political power with every other, as happens in a competitive
market in economics. This of course is afar of today's political reality.
Today in order to influence political decision, we   have to belong to some

group structure : to a labor union, to an industrial association, to a regional association, so the individual is mediated by many groups .

Maintaining an economic terminology, we may say that we have an oligopoly structure power .

With the word "Establishment" I mean the workable coalition between the power oligopolists in a country .

This is a coalition that should not be identified with only one productive class . This coalition may be stable or unstable. There may be a disturbance from a member of the coalition that wants more power for himself or there may be a disturbance from a person excluded from the coalition, that wants to join the coalition and so alter the balance of power.

It becomes more or less inevitable that the plan incorporates the values of the establishment.

In general if the coalition is somehow stable or reasonably balanced, I think that the plan itself will be a very mild plan. There will not be very much change proposed. The plan will not propose any radical change in any direction.

When a plan seems to alter the direction of a country's progress, then this must mean either one of two things :

1 - That a change has taken place in the balance of power within the coalition, or

2 - That the government in question is taking a risk, that is to say it is proposing a plan that is forward-looking in a certain direction and it is moving to challenge itself the structure and impose a change on the coalition of power in question.

A. Papandreou

One thing is clear . A plan that departs from a routine extension of the activities of a nation is associated, either with a historically given change in the structure of power of this establishment , or with an intended change from the part of the government that proposes the plan.

So the plan, it seems to me, must be seen not only as a scientific document , but must be seen also as a political document and as a clue to the state of affairs of the country ; the political balance of power, whether there exists a crisis or not, a forward movement or not.

The plan is the focus of political and social conflict.

Stampa: Editoriale Grafica - Roma - Tel. 5890154